CONTENTS

This book is divided into ten sections, each containing a variety of specimens. This should help in planning a trip for just a weekend or for a week. There are maps showing locations of the mineral samples. Special thanks to Bill Marshall and B.C. Forest Service for the use of the maps. *These maps are not to scale!*

ROCKS 8 NORTH KAMLOOPS TO LITTLE FORT

ROCKS 9 NORTH B.C.

ROCKS 10 KOOTENAYS

Gold panning the Fraser River.

Quartz crystal.

Fraser River jade.

OPPOSITE:
Rock formations on Deadman Creek.

INTRODUCTION

This second edition of *Gem Trails in British Columbia* has been produced to provide rockhounds with an updated but still only partial list of known mineral and fossil locations. Many of the locations from the first book are in this new edition and continue to be productive collection sites. Some sites have been omitted for various reasons: they have been claimed, access is no longer allowed, roads have changed, or very few gems are left. New areas have been added including Whistler–Lillooet and the Kootenays.

The information provided in this book has come from personal visits, other publications and from discussions with other rockhounds. All of the areas listed in this book have been visited by myself, my brother, or my sister. The locations described should be used as an indication of where to start your search and what to look for. The mention of an area in this book *by no means indicates permission*, so, permission to cross or dig on private land must be obtained. Also, in active mining areas, arrangements with the operators must be made in advance. Most of the areas listed are on public lands.

Some of the most useful companions for rockhounding these areas are the *B.C. Recreational Atlas* and forestry maps showing the lakes, streams, and campsites near each location. These maps are available for a nominal fee at the forestry stations in each area. If you decide to go into an area near a town or city, be sure to get in touch with the local rock club. They can be most helpful and more than happy to tell you of any recent discoveries, as well as conditions in the area you are going to visit.

site 1 WHISTLER

▸ map page 9

black tusk crystals

From the town of Squamish, at the A&W, head north on Highway 99 for 44 km. Turn on the road for the Whistler bungee jump. Follow this road past the government forest camp sites and the bungee jump for 5 km and turn right on the road heading toward Daisy Lake. This is the first road to the right after the bungee jump. Follow this road for 1.3 km where the crystals are in the road cut on the left. The next location is 1 km past here. There are more crystal locations around here so watch for the diggings of others. The crystals here are very nice. Some are clear and others are coated with a green coating of "pixi dust" (possibly micaceous chlorite). Pyrite has also been found back on the road that the bungee jump is on. *[Site information courtesy of Greg Peterson]*

Site 1, second crystal location.

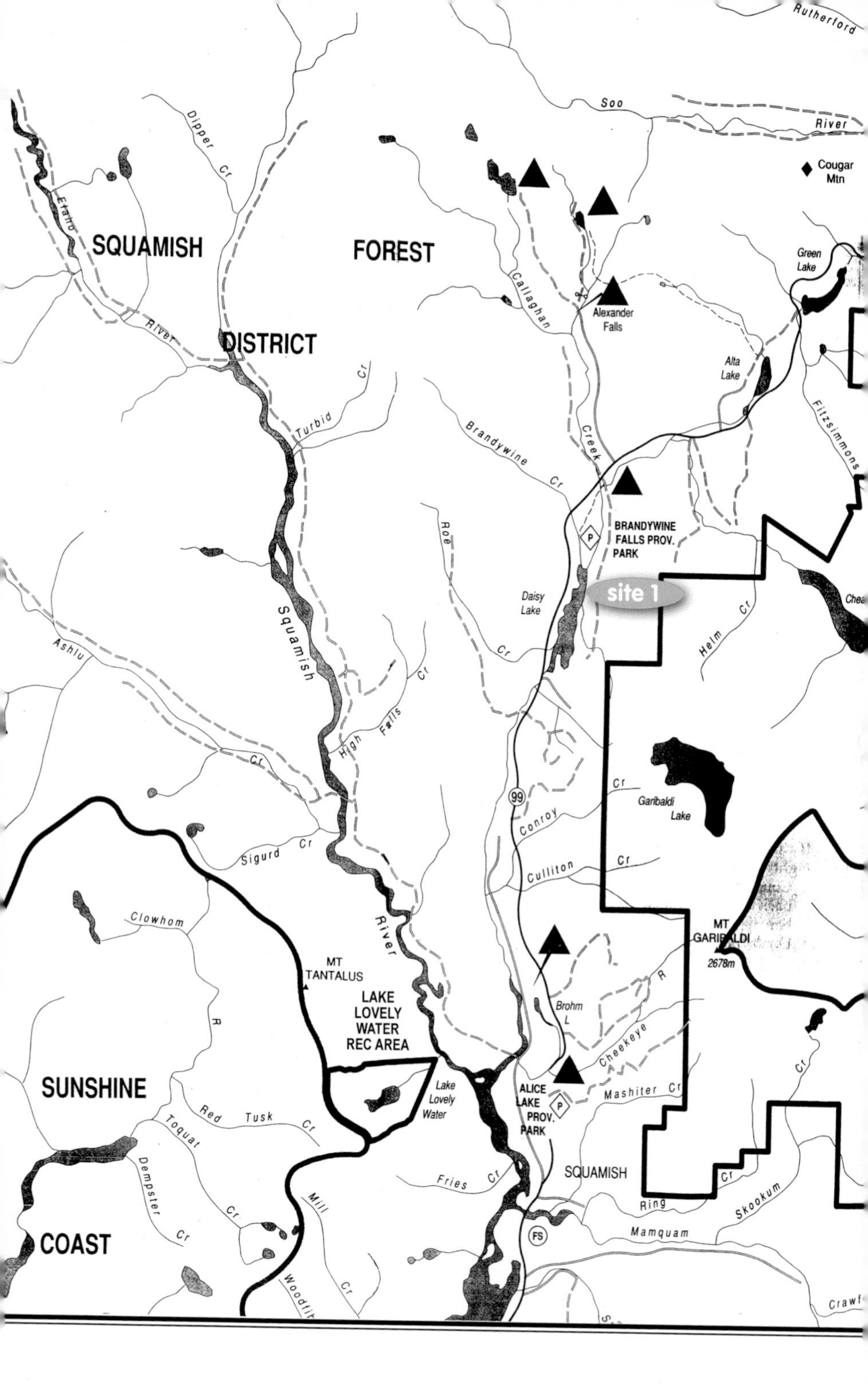
Rutherford
Soo
River
Cougar Mtn
Dipper Cr
Elaho
SQUAMISH
FOREST
DISTRICT
Callaghan
Alexander Falls
Green Lake
Alta Lake
River
Turbid
Cr
Brandywine Cr
Creek
Fitzsimmons
BRANDYWINE FALLS PROV. PARK
Roe
Daisy Lake
site 1
Cr
Helm
Cr
Ashlu
Squamish
High Falls
Cr
Cr
Garibaldi Lake
99
Conroy
Sigurd Cr
Culliton
Cr
Clowhom
River
MT GARIBALDI
2678m
MT TANTALUS
R
LAKE LOVELY WATER REC AREA
Brohm L
R
Cheekeye
Cr
SUNSHINE
Lake Lovely Water
ALICE LAKE PROV. PARK
Mashiter Cr
Red Tusk Cr
Toquat
Dempster
Cr
Cr
Mill
Fries Cr
SQUAMISH
Ring
Skookum
Cr
COAST
FS
Mamquam
Woodfibre
Cr
Crawford

site 2 LILLOOET

▸ map page 11

jade, jasper, serpentine, soapstone

The area around the town of Lillooet is a great place to find and learn about jade. Jade is a major attraction of the area and magnificent samples can be seen throughout the town. Jade samples can be found in front of most public buildings and businesses.

It is a large area to check out with many trips you can go on. The jade museum is a good place to start your search. The manager will give you lots of good advice which can help you in your search.

The best area to look for jade is along Bridge River Road. You can get there by traveling through the town and turning left on Bridge Lake Road.

As you travel around the lakes look for jade in ditches and on roadsides. Turn up the road to Marshall Lake; this is the best road to find jade in the ditches. There are jade mines in the area, so please respect the property of others.

There is also a lot of serpentine which will take a great polish. It is quite soft, making it easy to work with.

For jaspers, in town you can hunt the beach at the Cayoos Creek area which is also a panning reserve. Check with the visitor information center in town for directions. All of the Fraser River and most of the surrounding creeks will have claims on them, but the many forest service roads will be great places to search.

Boulders along the Fraser River.

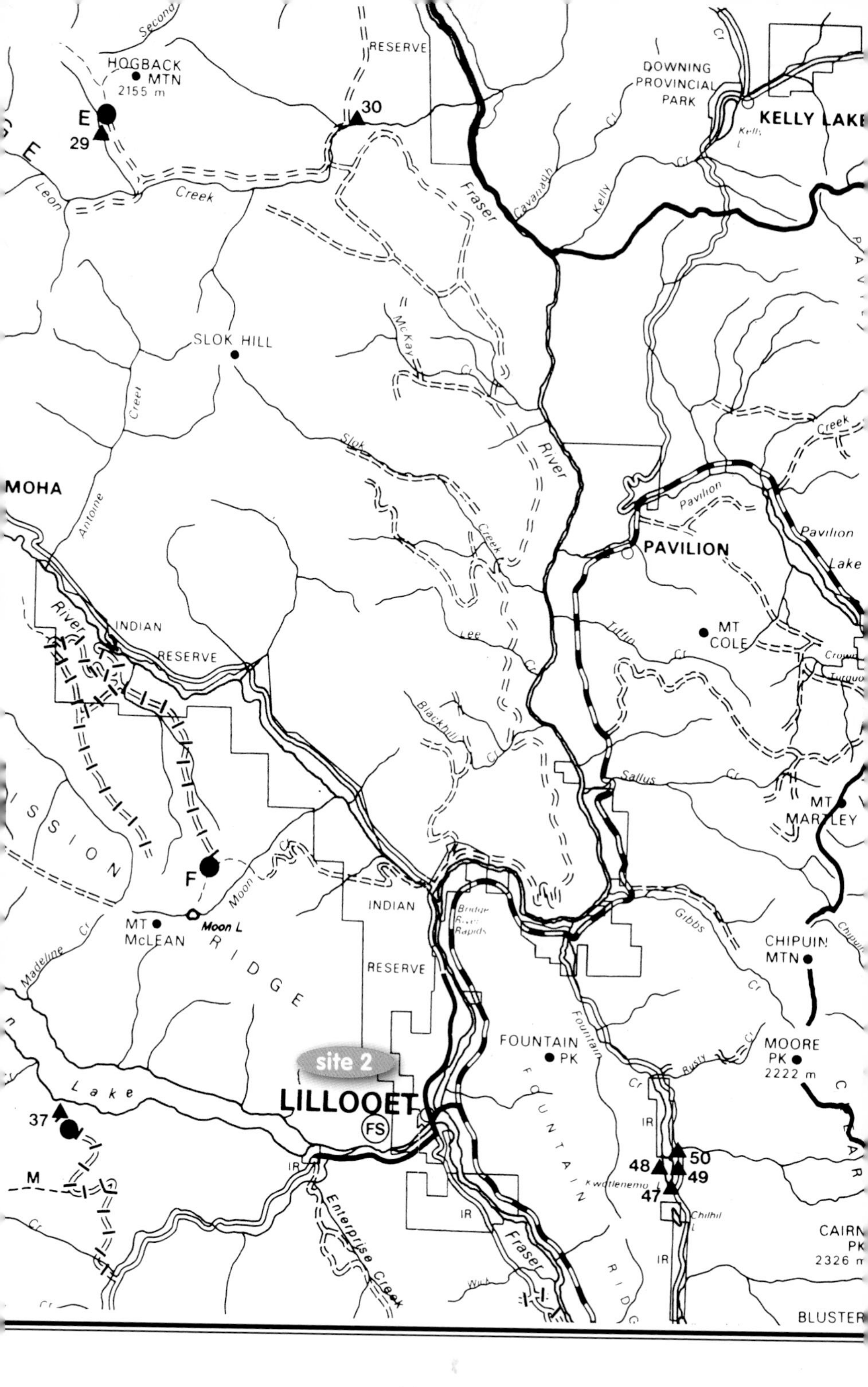
HOGBACK MTN
2155 m
E
29
30
RESERVE
DOWNING PROVINCIAL PARK
KELLY LAKE
Leon
Creek
Fraser
Cavanagh
Kelly
SLOK HILL
McKay
Creek
MOHA
Antoine
River
INDIAN
RESERVE
Slok
Creek
PAVILION
Pavilion
Lake
MT COLE
Tiffin
Lee
Blackhull
Sallus
MT MARTLEY
F
Moon L
MT McLEAN
Moon
RIDGE
INDIAN
RESERVE
Bridge River Rapids
Gibbs
CHIPUIN MTN
Madeline
FOUNTAIN PK
Fountain
FOUNTAIN
MOORE PK
2222 m
site 2
Lake
LILLOOET
FS
37
IR
Enterprise Creek
48
50
49
47
Kwotlenemo
Chilhil
CAIRN PK
2326 m
Fraser
BLUSTER

site 3 FRASER RIVER

▸ map page 13

agate, rhodonite, jasper

The gravels on the south side of the Fraser River from Chilliwack to Hope have produced a variety of mineral specimens suitable for gem stones. There are too many areas to name them all, but here are a few of the better ones. One of the easy access areas can be found under the Agassiz Bridge. To get to these locations take Highway 1 from Vancouver and turn off at the Agassiz, Harrison Hot Springs exit. This is Highway 9. As you approach the bridge, there is a side road on the right. This road is the Rosedale Ferry Road. It will run parallel to and then cross under the bridge where there are many pullovers and turnoffs to the river. Some of these turnoffs take you to very large rock bars. Search for agates by walking into the sun, and it is best after a rain, when the rocks are their cleanest. Petrified wood and bones can be hard to recognize, so proceed slowly. The jade and rhodonite can be quite large, so don't ignore the areas that have larger rocks. The waters of the Fraser River are at their lowest in the early spring, when the snow is still in the mountains. Keep fires small and, as always, pack out your garbage.

site 4 FRASER RIVER NORTH

▸ map page 13

agate, jade

The fishing bars on the north side of the Fraser River between Dewdney and Deroche have many good collecting areas to explore. One of the best entrances to the bars is to take Highway 7 from Vancouver, through Mission, past Dewdney, and just before you cross the bridge into Deroche, turn right on Athey Road. Follow it to the end, about 2.8 km, where the pavement ends. They have installed gates, so from here you will have to walk. If you turn right you will get to Wing Dam Bar and if you turn left and take the first right you will be at Queens Bar. Both are good for agates and other small gem stones. The best time to search the bars for agate is on a sunny day after a recent rain. The best time

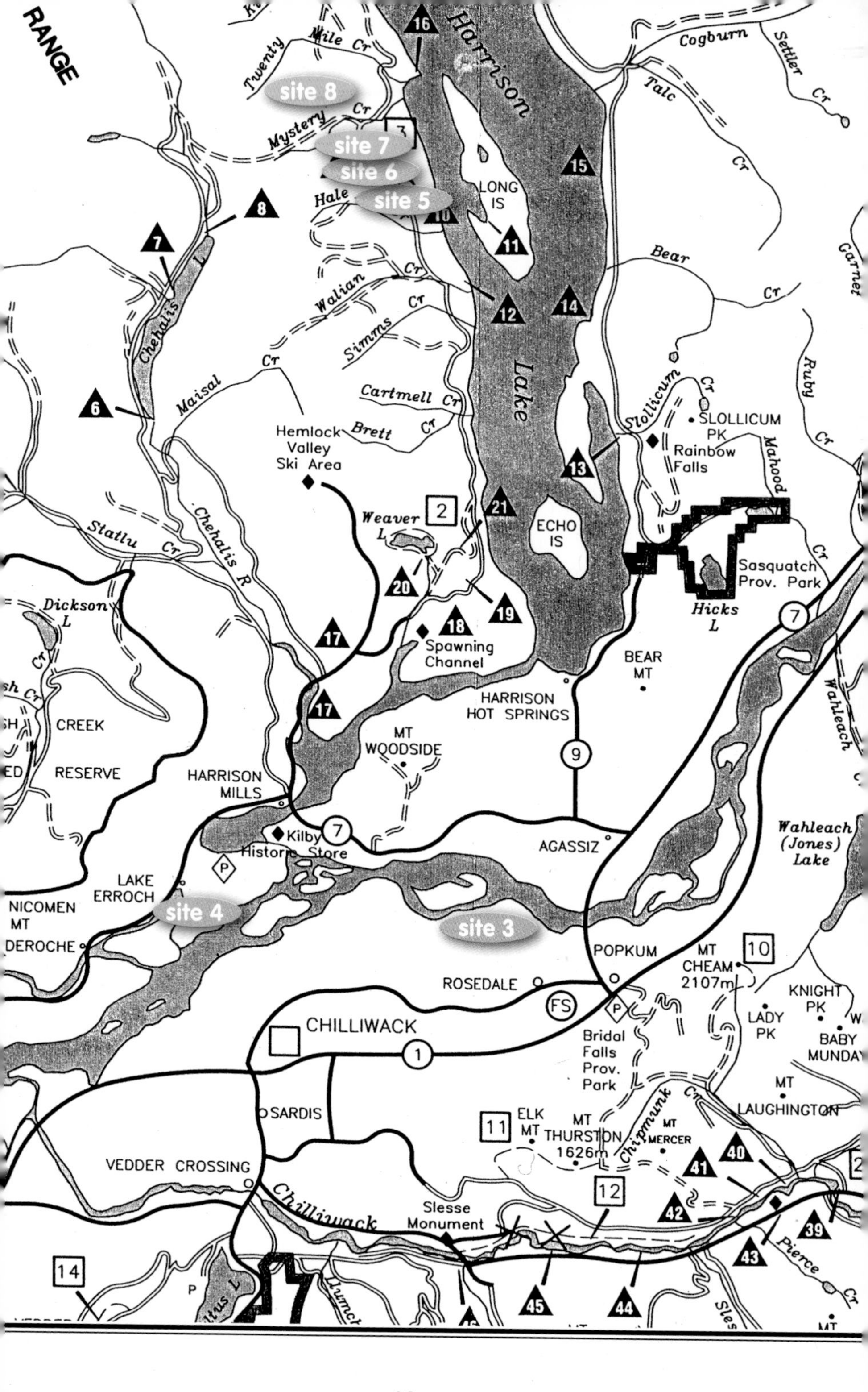

RANGE
Harrison Lake
site 8
site 7
site 6
site 5
site 4
site 3
LONG IS
ECHO IS
Hemlock Valley Ski Area
Weaver L
Spawning Channel
HARRISON HOT SPRINGS
BEAR MT
MT WOODSIDE
HARRISON MILLS
Kilby Historic Store
LAKE ERROCH
NICOMEN MT
DEROCHE
AGASSIZ
POPKUM
ROSEDALE
CHILLIWACK
SARDIS
VEDDER CROSSING
Slesse Monument
ELK MT
MT THURSTON 1626m
MT MERCER
MT CHEAM 2107m
LADY PK
KNIGHT PK
BABY MUNDAY
MT LAUGHINGTON
Bridal Falls Prov. Park
SLOLLICUM PK
Rainbow Falls
Sasquatch Prov. Park
Hicks L
Wahleach (Jones) Lake
Dickson L
CREEK
RESERVE
Chehalis L
Chehalis R
Statlu Cr
Maisal Cr
Walian Cr
Simms Cr
Cartmell Cr
Brett Cr
Mystery Cr
Twenty Mile Cr
Hale
Cogburn
Talc Cr
Settler Cr
Bear Cr
Garnet
Ruby Cr
Slollicum Cr
Mahood Cr
Wahleach
Chilliwack
Chipmunk Cr
Pierce Cr

of the year is December to February, because the river is at its lowest in early spring. The gravel bars are used by fisherman and rockhounders for camping, but you must pack out your garbage. Some people have not cleaned up, and this has led to the closure of some areas.

site 5 HALE CREEK

▸ map page 13

ammonite, fossils

From Mission, Maple Ridge take Highway 7 to the west side of Harrison Lake. Turn left on Morris Valley Road (at the Sasquatch Inn), proceed through the stop sign, and stay on the main road. At 35.3 km, just after Hale Creek, you will find ammonites in the road cut. The ammonites are in the bank of the road on the left-hand side. The rock is, I believe, basalt and quite dark in color. The ammonites in this area are great for displays but not for polishing. There should be more locations of these in the immediate area. A good place to start the search may be in the wash just before the creek or in the creek itself. The ammonites are easily broken, but you must keep the ditch of the road clear for drainage. In this area there are many good places to camp, including many forestry sites with tables and washrooms. Please respect this area and take out your garbage.

Hillside at Hale Creek.

site 6 WEST HARRISON

▸ map page 13

sealife fossils

From Mission, take Highway 7 to Morris Valley Road, on the west side of Harrison Lake (at the Sasquatch Inn). Turn left and proceed through the stop sign, and stay on the main road. Proceed for 37 km and in this road cut you will find a variety of fossils, some of which have been agatized. They are in the hard rock on the left-hand side of the road. Many of these fossils are interesting and may take a polish. Check the ditch below the fossils and also in the roadfill on each side to get samples. This will be easier than to attack the main host rock, as it is quite hard.

site 7 BODEE CLAIM

▸ map page 13

clam fossils

From Mission, take Highway 7 to Morris Valley Road, on the west side of Harrison Lake (at the Sasquatch Inn). Turn left and proceed through the stop sign, and stay on the main road. Around 39.2 km you will find many clam fossils. Some of the rocks containing the fossils are the size of cars, so be sure to take hammers and chisels to get the best samples. The many campsites in this area will make your stay quite comfortable and you may want to try fishing for your dinner in the many small lakes as well as in Harrison Lake itself. There are many fossils in this area of the lake, but this is a good place to start. A full, detailed map is available at the geological survey of Canada (Vancouver). This area has many fine lakefront forestry camps, but you must clean up after yourself, as there is no garbage pick-up. *[Site access courtesy of Dee and Bob Morgan]*

site 8 MYSTERY CREEK

▸ map page 13

fossils

From the Sasquatch Inn on Highway 7 east of Mission, turn left and follow the Morris Valley road for 41 km. This location is just past the other fossil sites. Turn left on Mystery Creek Road. At .3 km turn right and travel a short distance to the road split. Just past the split on both roads there are fossils, mainly clam. When they plowed a short road to the left, there were clam fossils that resembled butter clam shells. This road has now been deactivated.

Mystery Creek hillside just after the fork in the road.

site 9 EAST HARRISON LAKE

▸ map page 17

garnet

To get to the town of Harrison Hot Springs, take the Agassiz exit off Highway 1 at exit number 135. Go through Agassiz and follow the signs to Harrison Hot Springs. The first stop sign is at Lillooet Avenue. Turn right and follow the road around the lake. Keep right at the park until you see the gravel road on the left that is Harrison Lake east forest service road. Follow this road for 23 km. Here you will be at a log sorting station. After the station (1.1 km) keep left and continue to follow above the lake. After 10 km you will be at another sorting station. Continue to Pine Mainline Road which goes off to the left at 3.3 km. From here you will need a high-clearance, four-wheel-drive vehicle. Cross the river and pro-

East Harrison Lake garnet search.

site 9
ILLIWACK
FOREST
Tipella Cr
Stokke Cr
Tretheway Cr
Bremner Cr
Silver Cr
Clear Cr
Hornet Cr
MT URQUHAR
2100m
Davidson Cr
Kirkland Cr
Twenty Mile Cr
Mystery Cr
Cogburn
Settler Cr
Talc Cr
Harrison Lake
Winslow Cr
Hale Cr
LONG IS
Walian Cr
Simms Cr
Bear Cr
Garnet
Roaring Cr
Kenyon L
Blinch L
Chehalis L
Maisal Cr
Cartmell Cr
Brett Cr
Ruby
Slollicum Cr
SLOLLICUM PK
Rainbow Falls
Mahood
Hemlock Valley Ski Area
Salsbury L
Lost
Statlu Cr
Chehalis R
Weaver L
ECHO IS
SASQUATCH PROV. PARK
Hicks L
Dickson L
L W Norrish Cr
DAVIS LAKE PARK
MT ST BENEDICT
Cascade Cr
Cascade Creek Falls
NORRISH CREEK
WATERSHED RESERVE
Norrish
Spawning Channel
HARRISON HOT SPRINGS
BEAR MT
MT WOODSIDE
HARRISON MILLS
Kilby Historic Store
AGASSIZ
Wahleach Cr
Wahleach (Jones) Lake
LAKE ERROCH
NICOMEN MT
DEROCHE
DEWDNEY PK
DEWDNEY
POPKUM
MT CHEAM
2107m
ROSEDALE
KNIGHT PK
LADY

ceed up the hill for 2 km. Garnets are here in the road cut and in the field beside. There are many locations for garnets around here so travel slow and watch for the red dots. Travel 2.4 km then take the right fork in the road. A short distance from the road fork you will find garnets and graphite in the road cut. *[Site information courtesy of Mike Blampied]*

site 10 RUBY LOUGHEED CREEK FSR

▸ map page 19

jasper, garnet

From Hope, heading north on Highway 1, cross the Fraser River and travel to the junction of Highway 1 and Highway 7. From this junction head west on Highway 7 for 13 km and you will be at the Ruby Lougheed Creek forest service road. On this road you can search for garnet as float. The main source has not been found but you can find the garnets in the silvery schist host rock in different locations along the road.

site 11 GARNET CREEK FSR

▸ map page 19

garnet

From Hope, heading north on Highway 1, cross the Fraser River and travel to the junction of Highway 1 and Highway 7. From this junction head west on Highway 7 for 11.3 km and you will be at the Garnet Creek forest service road. As you travel this road watch for the schist that will contain the garnet. You can find garnets at the mouth of Garnet Creek where it enters Ruby Creek.

Turnoff at Highway 7.

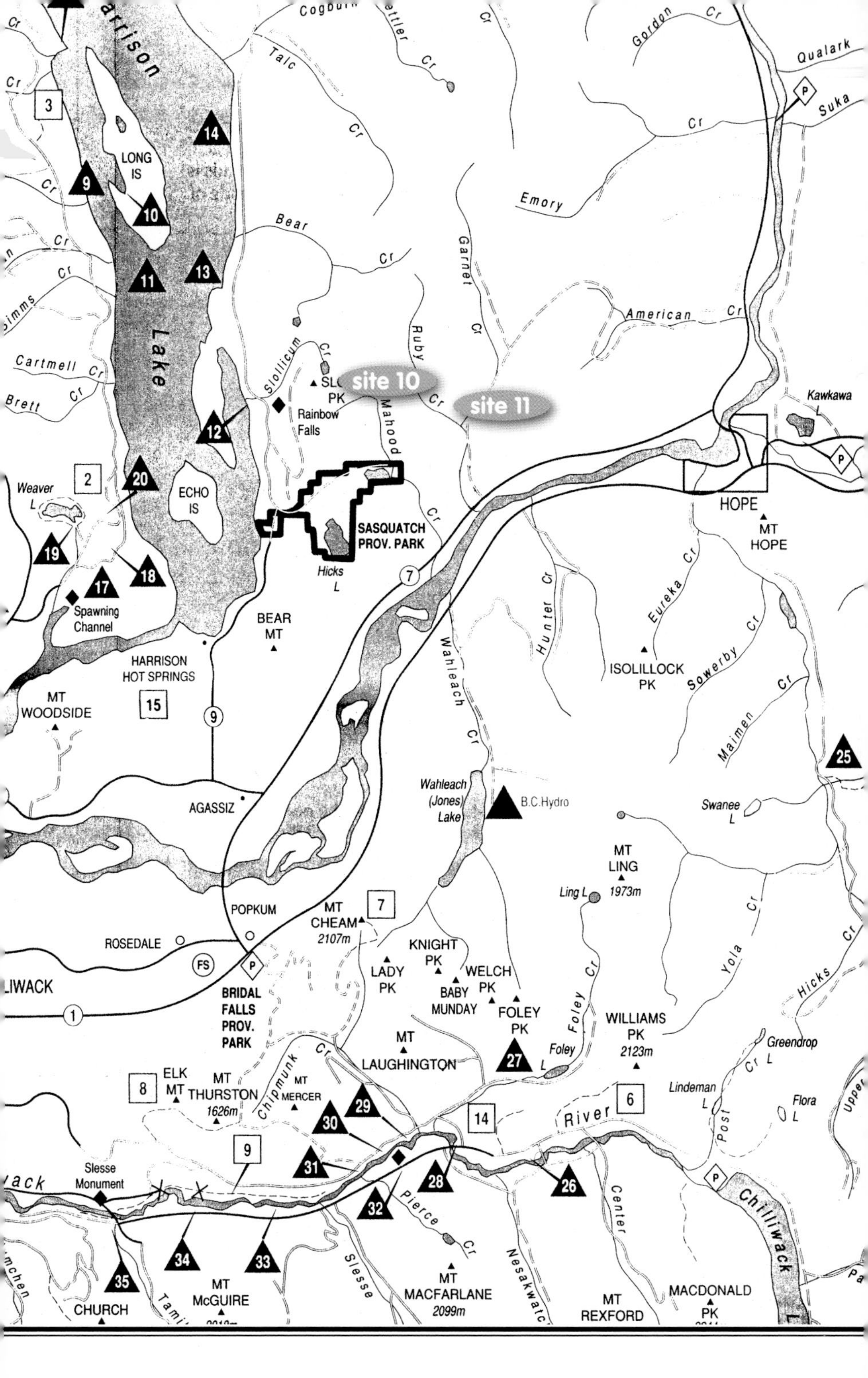

Harrison Lake
LONG IS
ECHO IS
Talc Cr
Bear Cr
Emory
Gordon Cr
Qualark
Suka Cr
Garnet Cr
Ruby
American Cr
Slollicum Cr
SLOLLICUM PK
site 10
site 11
Mahood Cr
Rainbow Falls
Kawkawa L
HOPE
MT HOPE
SASQUATCH PROV. PARK
Hicks L
Weaver L
Cartmell Cr
Brett Cr
Simms
Spawning Channel
HARRISON HOT SPRINGS
BEAR MT
MT WOODSIDE
Hunter Cr
Eureka Cr
ISOLILLOCK PK
Sowerby Cr
Maimen Cr
Wahleach Cr
Wahleach (Jones) Lake
B.C.Hydro
Swanee L
AGASSIZ
MT LING
1973m
Ling L
POPKUM
ROSEDALE
MT CHEAM
2107m
KNIGHT PK
LADY PK
WELCH PK
BABY MUNDAY
FOLEY PK
Foley Cr
Foley L
Yola Cr
Hicks Cr
WILLIAMS PK
2123m
Greendrop L
Lindeman L
Flora L
Post Cr
BRIDAL FALLS PROV. PARK
MT LAUGHINGTON
ELK MT
MT THURSTON
1626m
Chipmunk Cr
MT MERCER
River
Chilliwack L
Slesse Monument
Pierce Cr
Center
Slesse
Nesakwatch
MT MACFARLANE
2099m
MT McGUIRE
MT REXFORD
MACDONALD PK
CHURCH

site 12 HYDRO STATION

▸ map page 21

quartz crystals, epidote

Proceed on Highway 1 to the hydro substation 8.6 km north of the Agassiz turnoff. Just .8 km before the station there is a road leading up the hill and this is a good place to park. You can park on the road just past the hydro station. The crystals are .5 km up the creek. Check the creek as you go, but you will have to use the road in some sections, as the creek bank is very steep. They have returned the road to its natural state. Flagging has been installed to assist you. If you follow the old road clearings and head toward the water flume clean out, you will come to the creek. The crystals are up 200 metres and to the left 100 metres. There is a faint trail to follow; it starts about 40 metres up the hill. The climb is steep, but take your time and follow the trail to a creek bed that is sometimes dry; it is more of an overflow. It is in this overflow that the crystals are found. If you have time, be sure to look around for more deposits as small samples have been found in other areas near this location. One of the areas to check is by the hydro station, where they deposited the rocks removed from the tunnel dug for the water used at the hydro station.

Hydro station on Highway 1.

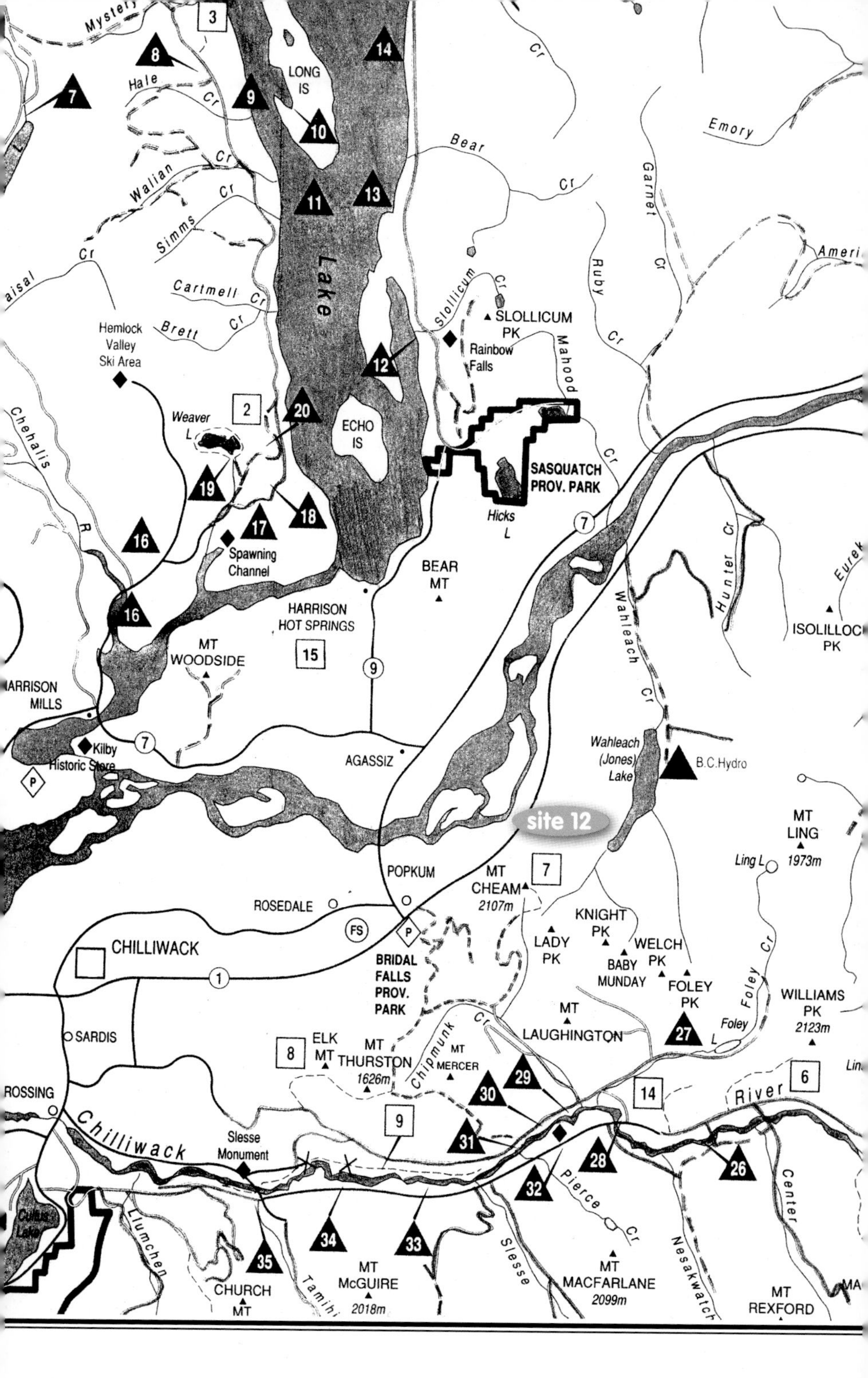
LONG IS
Lake
ECHO IS
Hemlock Valley Ski Area
Weaver L
Spawning Channel
SASQUATCH PROV. PARK
Hicks L
SLOLLICUM PK
Rainbow Falls
BEAR MT
HARRISON HOT SPRINGS
MT WOODSIDE
HARRISON MILLS
Kilby Historic Store
AGASSIZ
Wahleach (Jones) Lake
B.C.Hydro
site 12
POPKUM
ROSEDALE
MT CHEAM 2107m
CHILLIWACK
BRIDAL FALLS PROV. PARK
LADY PK
KNIGHT PK
WELCH PK
BABY MUNDAY
FOLEY PK
MT LING 1973m
Ling L
WILLIAMS PK 2123m
ISOLILLOC PK
MT LAUGHINGTON
Foley L
SARDIS
ELK MT
MT THURSTON 1626m
MT MERCER
Chilliwack
River
Slesse Monument
Cultus Lake
CHURCH MT
MT McGUIRE 2018m
MT MACFARLANE 2099m
MT REXFORD

site 13 SKAGIT

▸ map page 23

axinite, quartz crystals

Proceed along Highway 3 from Hope, heading toward Princeton. At 34.6 km you will come to Skagit Bluffs. Just as you enter the bluffs, you will see a small rock standing on the right-hand side of the road. This is the best location for the quarter-inch axinite crystals of violet and black, as well as quartz crystals with unique green inclusions (up to two inches long and in clusters). They are in the road cut and as well as over the bank. Each year they seem to do some work on this part of the road, so be sure to check other locations. This is in Manning Park area and the status is unknown. You should check at the park office half way between Hope and Princeton. This is about 69 km from either Hope or Princeton. Be sure to check with the local weather station in the winter as snow can be present from November to February.

There are crystals at the base of this outcrop.

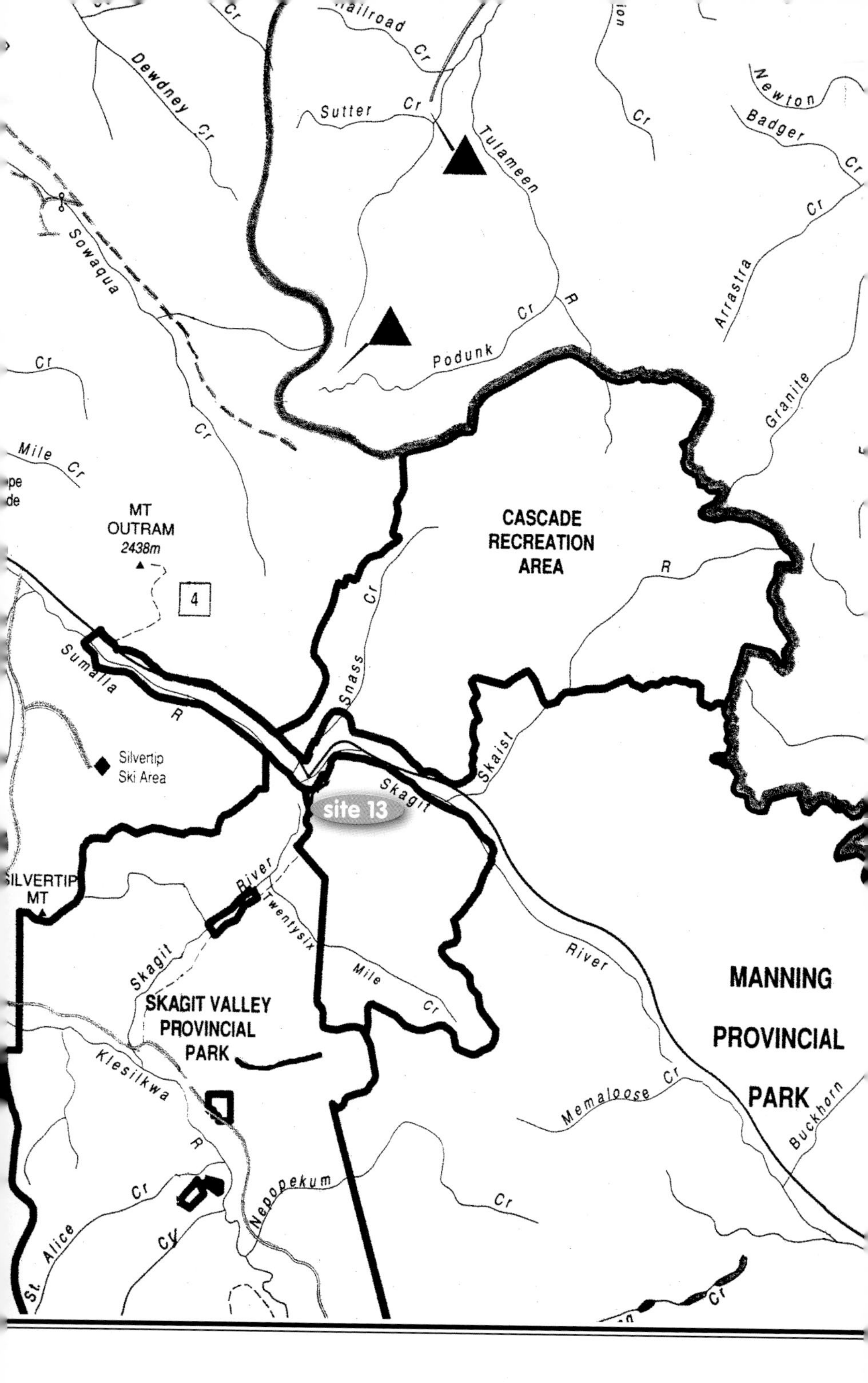
Railroad Cr
Dewdney Cr
Sutter Cr
Tulameen R
Newton
Badger Cr
Cr
Sowaqua Cr
Arrastra Cr
Podunk Cr
Granite
Cr
Mile Cr
MT OUTRAM
2438m
4
CASCADE RECREATION AREA
R
Snass Cr
Sumallo R
Silvertip Ski Area
Skaist
Skagit
site 13
SILVERTIP MT
River
Twentysix Mile Cr
Skagit
River
MANNING PROVINCIAL PARK
SKAGIT VALLEY PROVINCIAL PARK
Klesilkwa R
Memaloose Cr
Buckhorn
Nepopekum
Cr
St. Alice Cr
Cr
Cr

site 14 SUNDAY SUMMIT

▸ map page 25

petrified wood, zeolites

Proceed up Highway 3 to Sunday Summit which is 124 km from Hope. Just before the summit and on the east side of the road, on the bluffs and in the talus, you will find the wood. Good material is scarce, but when I was last there they were logging the area and this should open up more bluffs and road cuts making the wood easier to find. It will be easier to go around on the logging road to search for the material than to trek through the woods. With the new cuts and a bit of luck, you should be able to find more sights as well as some new materials. There is plenty of free wilderness camping with wood and water. The best campsite in this area is at Copper Creek, where you will find tables, washrooms and good water. The site is located west of the summit at the bottom of the hill on the Copper Creek Road. This area is at the 5,000-foot level, so be sure to check that the snow has not covered the sites, particularly from October to April.

Sunday Summit turnoff.

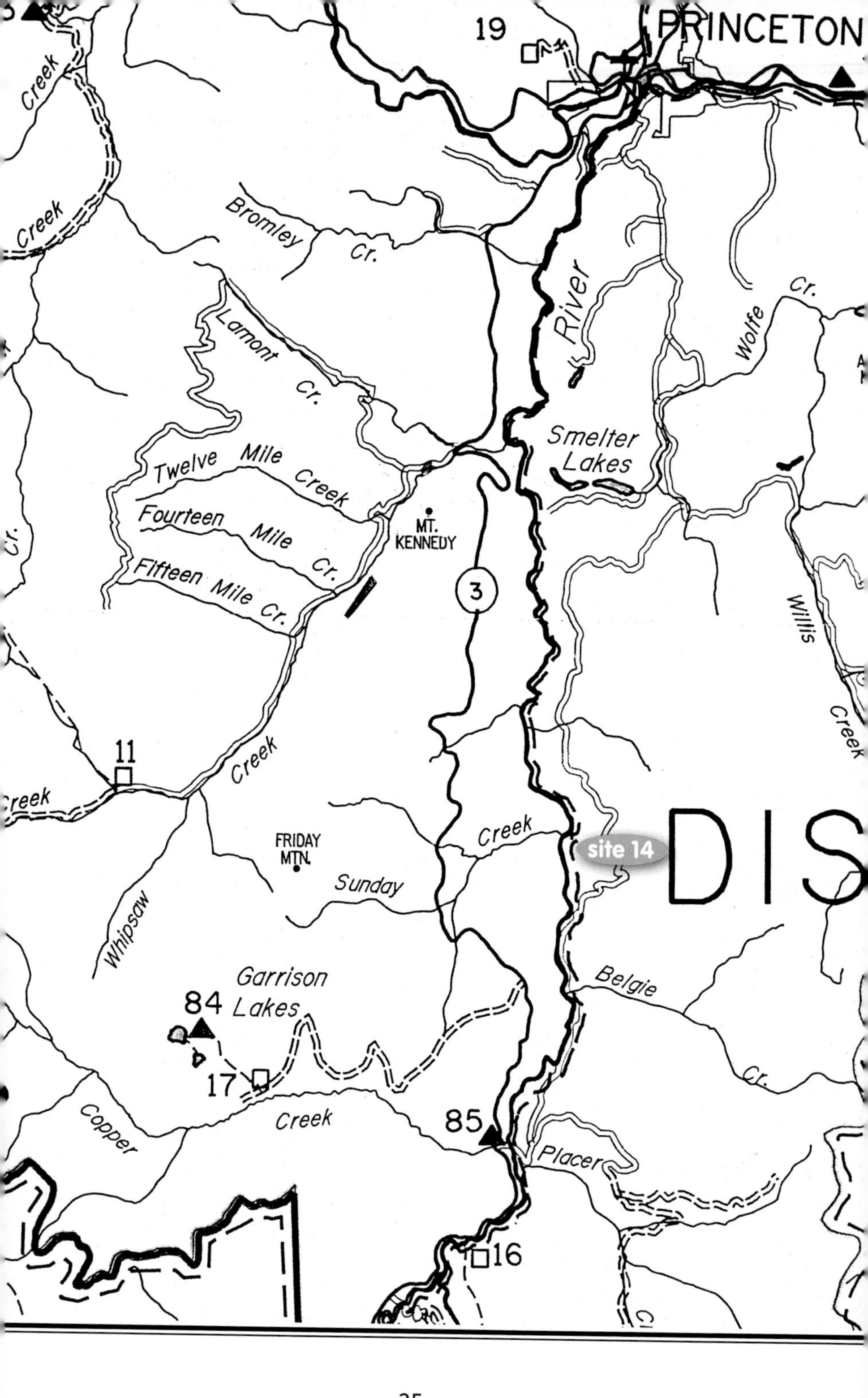
PRINCETON
19
Creek
Creek
Bromley
Cr.
River
Cr.
Wolfe
Lamont
Cr.
Smelter
Lakes
Twelve Mile Creek
Fourteen Mile Cr.
Fifteen Mile Cr.
MT.
KENNEDY
3
Willis
Creek
11
Creek
reek
Creek
FRIDAY
MTN.
Sunday
site 14
DIS
Whipsaw
Garrison
Lakes
84
17
Belgie
Cr.
Copper
Creek
85
Placer
16

site 15 WHIPSAW FSR 1

▸ map page 27

amber

On Highway 3, the Hope–Princeton Highway, take the Whipsaw Creek forest service road. This road is about 13 km west of Princeton. Proceed just past the cattle guard and park. Follow the fence back toward the highway and you will see an old road heading down to the creek. Just up from where the road meets the creek, there is a showing of coal and in this coal you will find amber. There are more coal locations in this area so be sure to check around. Please be sure to keep back from the creek as it is a water source for the cattle and local animals. At the mouth of Whipsaw Creek there is more coal with some fish fossils in it.

To see the many types of rocks and fossils that are in the Princeton area, be sure to stop in at the Princeton Museum, where they have a lot of them on display. The museum recently received a very large and good collection from a well-known collector.

Parking at Whipsaw site.

Martins Lake
PRINCETON
19
7
Creek
Creek
Bromley
Cr.
River
Cr.
Wolfe
Lamont
Cr.
site 15
site 16
Smelter
Lakes
Twelve Mile Creek
Fourteen Mile Cr.
Fifteen Mile Cr.
MT.
KENNEDY
3
Willis
Creek
11
Creek
eek
Creek
FRIDAY
MTN.
Sunday
DIS
Whipsaw
Garrison
84 Lakes
Belgie
Cr.
17
Copper
Creek
85
Placer
16

site 16 WHIPSAW FSR 2

fossils

▸ map page 27

Take the Whipsaw Creek forest service road from Highway 3, which is about 13 km west of Princeton. The area around Whipsaw Creek has many different types of rocks and collectables so keep your eyes open for other types of rocks.

The fossils in this area are the same as the ones near Coalmont. The fossils in the solid sandstone are more difficult to find, but they are both quite hard, making them more desirable. The sandstone I found the best for displays and strength was a beige color. Again, check at the museum in Princeton to see samples of the fossils.

Leaf fossils found on Whipsaw Creek forest service road.

site 17 VERMILLION BLUFFS

▸ map page 30

agate, common opal, petrified wood

This is a great area for a scenic walk, with the potential for some good finds. The gem stones are found just west of Princeton at Vermillion Bluffs. At Princeton Irly Bird turn right and continue to your right down Burton and Grandby Avenues. Park and walk through the old tunnel. Just after the tunnel there is the old bridge, cross it and follow the tracks for 1.8 km. The gem stones are in the railway cut, and the cliffs above the cut. Also be sure to check the river on both sides. There are some fossils in the railroad cut just after you cross the bridge. You can also access this area from the road going from Princeton to Coalmont. On this road the river is on your left and you will have to watch for a location that has easy access. To get to the river where the best locations are, it is a tougher hike from this approach, but a shorter hike. The gem stones are in the talus and in the high cliffs. Across the river at McCormick Flats there were agate and quartz crystals reported, but I could find no access. Some of the people from the Abbotsford rock and gem club had talked to the local ranchers and were allowed access. As always, be sure to take out all your garbage.

Overlooking Vermillion Bluffs from Highway 3.

Hillside of the Coalmont fossil collecting site.

site 18 COALMONT ROAD

▸ map page 30

fossils

From the bridge in town, go west of Princeton on the Coalmont Road for 6 km. On the right side of the road in the yellow shale, you will find leaf and other organic fossils. The shale is somewhat soft, and care should be taken to insure the fossils remain complete. When my brother was there a fish fossil was found, but I wasn't as fortunate. They recently repaired the road and used the fossil-bearing rock as fill in the area. As you drive through the area, watch for the yellow shale and check each location. To see the many types of rocks and fossils that are in the Princeton area, be sure to stop in at the Princeton museum, where they have a lot of them on display.

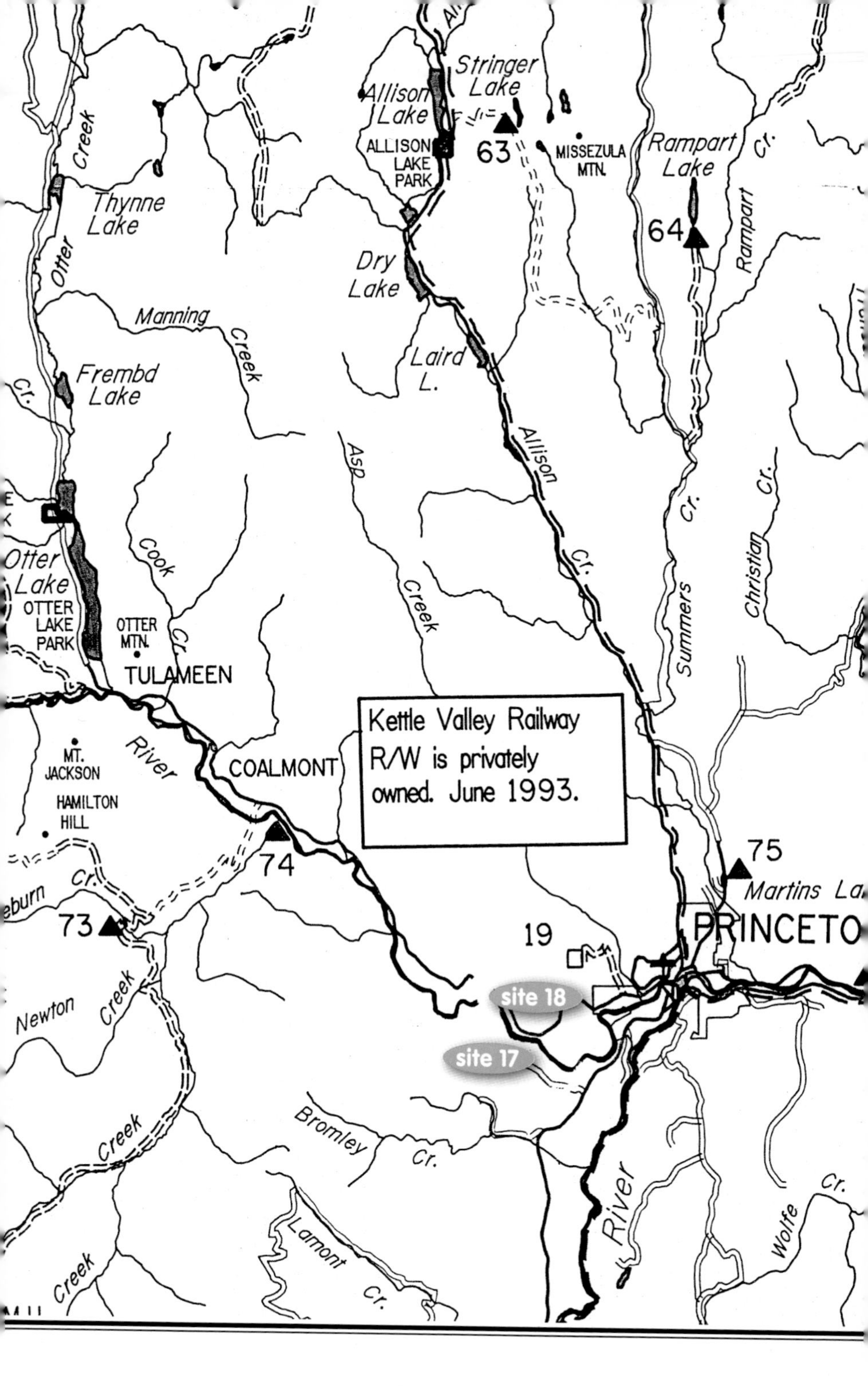
Stringer Lake
Allison Lake
ALLISON LAKE PARK
63
MISSEZULA MTN.
Rampart Lake
Cr.
Rampart
64
Creek
Thynne Lake
Otter
Dry Lake
Manning
Creek
Laird L.
Frembd Lake
Cr.
Allison
Asp
Cr.
Summers
Christian
Cr.
Cook
Otter Lake
OTTER LAKE PARK
OTTER MTN.
Cr.
Creek
TULAMEEN
Kettle Valley Railway R/W is privately owned. June 1993.
MT. JACKSON
River
COALMONT
HAMILTON HILL
75
74
Martins La
Cr.
73
19
PRINCETO
site 18
Newton
Creek
site 17
Bromley
Cr.
Creek
River
Cr.
Wolfe
Lamont
Cr.
Creek

site 19 BLAKEBURN GHOST TOWN ▸ map page 33

amber, fossils, gold

The ghost town of Blakeburn offers some great opportunities to see town remains and may be a good spot for a metal detector. To get there, continue straight through Coalmont instead of heading toward Tulameen. Cross the bridge and continue for 1 km and turn right at the remains of Granite City. Here at Granite City, which was quite large in its time, you can see the remains of some of the cabins. There is an old story that one of the miners in the area buried a large bucket of platinum near the town and it has never been found. Stay on the main logging road for 7 km and the old town site of Blakeburn is over the bank on the left. There are at least two roads into the large wall of coal. Camping in this area is very good, and one of the best campsites is just past the ghost town of Granite City. This site is at the junction of the two rivers; there are toilets and picnic tables.

Blakeburn wall of coal containing amber and fossils.

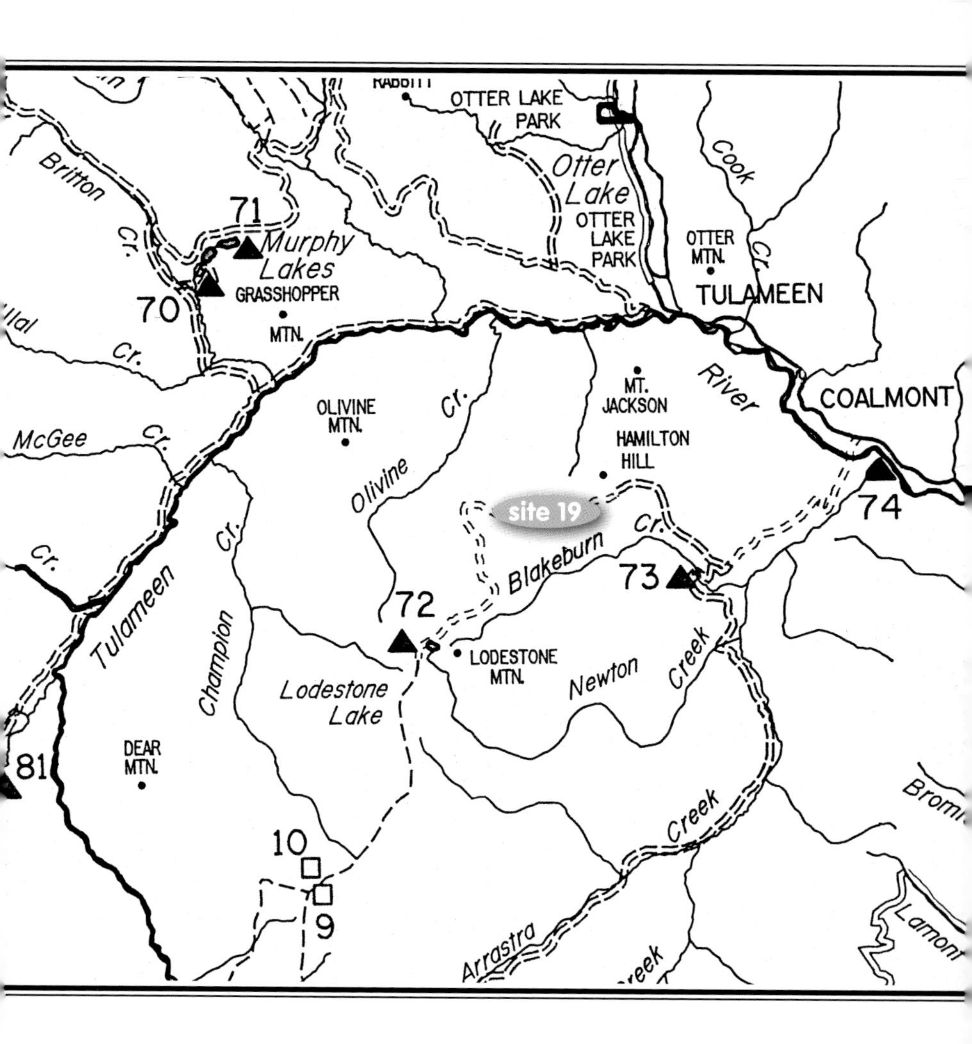
OTTER LAKE
PARK
Otter
Lake
OTTER
LAKE
PARK
Cook
Cr.
OTTER
MTN.
TULAMEEN
Britton
Cr.
71
Murphy
Lakes
70
GRASSHOPPER
MTN.
Cr.
McGee
Cr.
OLIVINE
MTN.
Cr.
Olivine
MT.
JACKSON
River
COALMONT
HAMILTON
HILL
site 19
Cr.
Blakeburn
74
73
72
Tulameen
Champion
Cr.
LODESTONE
MTN.
Newton
Creek
Lodestone
Lake
DEAR
MTN.
81
10
9
Creek
Arrastra
Lamont

LOCAL CREEKS AND RIVERS

placer gold

Nuggets and coarse gold were found in Granite and Lockie Creeks, as well as the Similkameen River. With the gold at Granite Creek you may also get some mercury in your pan. There are many small creeks in the area, so be sure to take a pan with you. To get a better idea where the gold came from, check at the local museum for the old town sites and mines, and be sure to watch for claims and active areas. The campsite I find the best in the Coalmont area, is the free forestry camp just after Granite City. There are washrooms, tables and water from the rivers. It is best to bring in drinking water, because of all the mining in the area. The bank robber Billy Miner lived for a short time in Princeton, and was thought to have hid some money in the hills east of town.

Collecting in the winter in local creeks and rivers.

site 20 DEWDNEY CREEK

▸ map page 36

serpentine, jade, soapstone, shmoos

From Hope head north on Highway 5 going toward Merritt. The best location for the serpentine that I have found is at, and around, the Carolin mine turn-off. They have used the serpentine for fill on the riverbanks to keep the high water from overflowing. When searching the road or riverbanks, look for the telltale signs of green rock. In these locations be sure not to damage the protective walls that contain the river at high water. Also, in this area there are two sites that the rock was removed from. The first location is half a mile south of the Carolin mine turnoff, on the right-hand side of the Coquihalla Highway. At this location you are well off the road and there are good campsites, but be sure the pit is not being used. It is 16 km from the location where Highway 3 turns off. The second location is a private pit east of the Carolin turnoff. On the no trespassing sign there is a phone number to call for permission to enter. Please respect private property and call.

Serpentine pit on the south side of Highway 5.

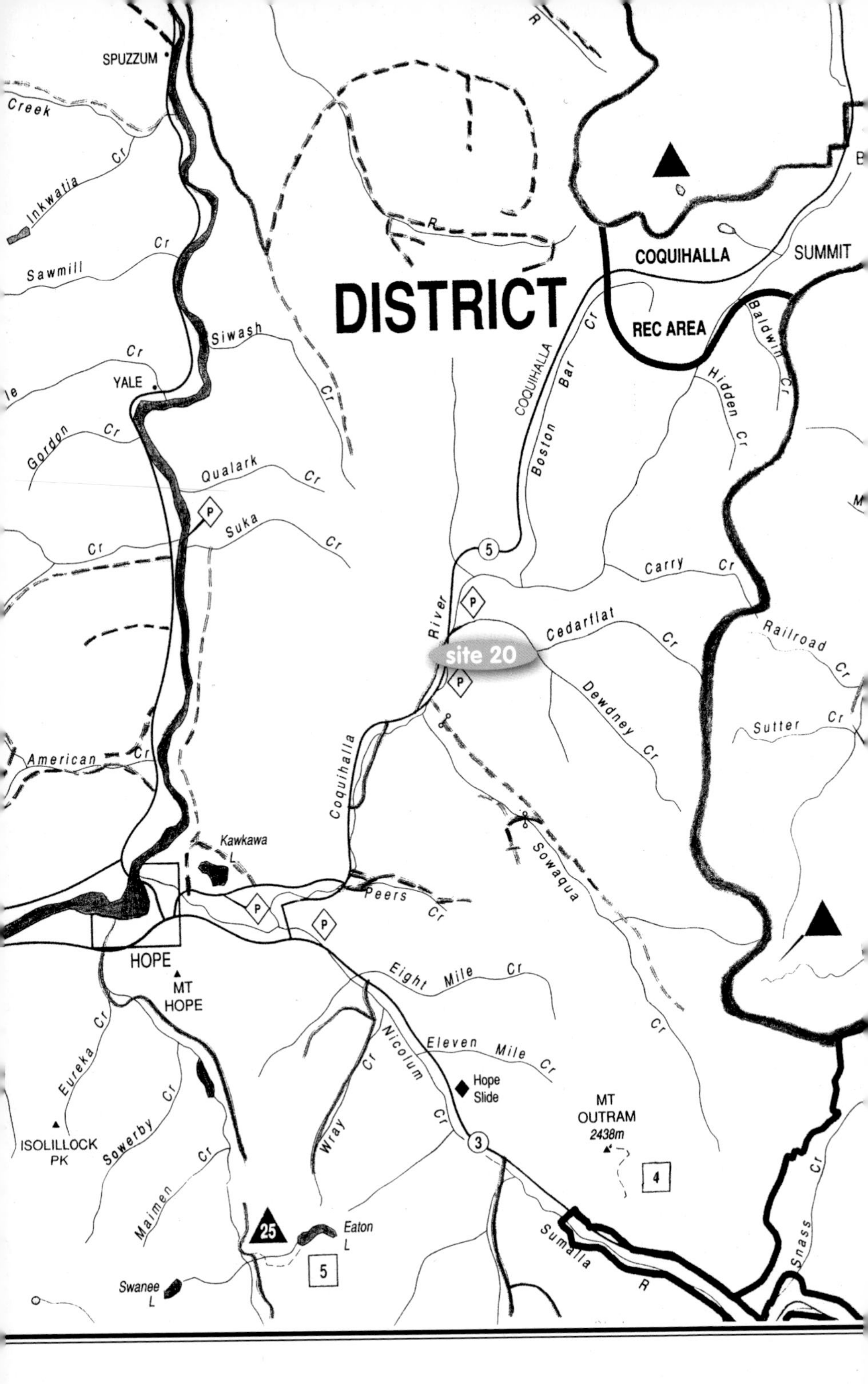
SPUZZUM
Creek
Cr
Inkwatia
Cr
Sawmill
Siwash
Cr
YALE
Gordon
Cr
Qualark
Cr
Suka
Cr
Cr
American
Cr
Kawkawa
L
HOPE
MT
HOPE
Eureka
Cr
ISOLILLOCK
PK
Sowerby
Cr
Maimen
Cr
25
Eaton
L
5
Swanee
L
DISTRICT
R
R
Cr
COQUIHALLA
SUMMIT
REC AREA
Baldwin
Cr
Hidden
Cr
COQUIHALLA
Boston
Bar
Cr
5
Carry
Cr
River
site 20
Cedarflat
Cr
Railroad
Cr
Dewdney
Cr
Sutter
Cr
Coquihalla
Sowaqua
Cr
Peers
Cr
Eight
Mile
Cr
Nicolum
Cr
Eleven
Mile
Cr
Hope
Slide
Wray
Cr
3
MT
OUTRAM
2438m
4
Sumalla
R
Snass
Cr

site 21 MINE CREEK ROAD

▸ map page 37

pyrite, bornite

The mine dumps containing pyrite and bornite.

From Merritt take Highway 5 heading toward Hope. At exit 231, which is 56 km from Merritt, exit at Mine Creek Road. As you come to the stop sign, turn to the right. Go up the hill and where the pavement ends you will be on July Creek Road. Follow this for only .4 km and you will see the Keystone Mine on your left and the dumps on your right. For the best results scrape the dumps and watch for the pyrite as you go. If you don't find any in your first attempt, watch for small amounts of pyrite or bornite and dig above them. Also, if you go back to the exit and turn left, you will find the same type of material with some pyrite in it.

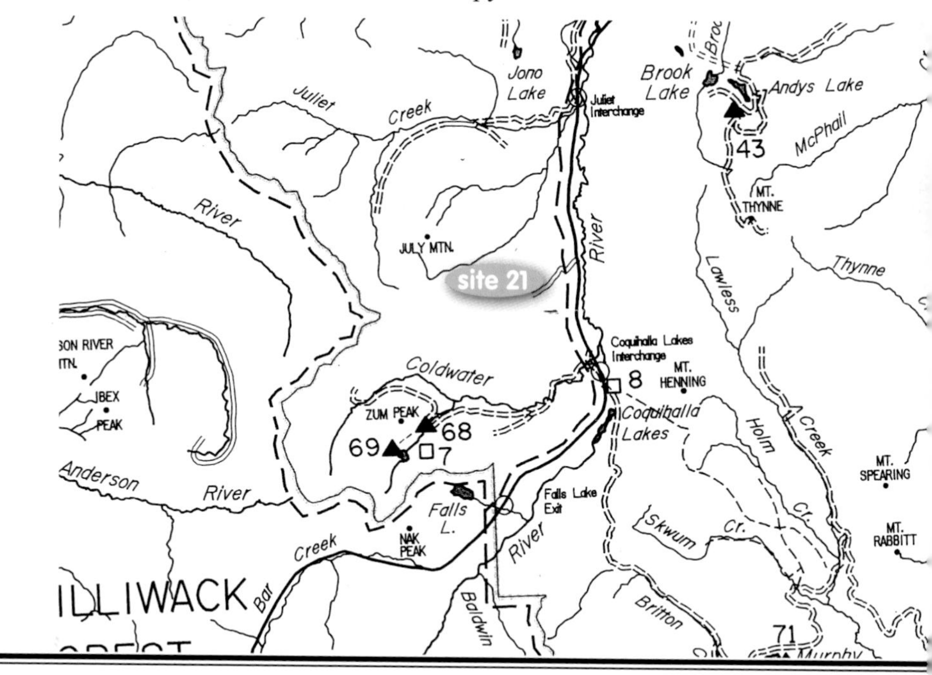

site 22 HOPE

▸ map page 39

jade, agate, jasper, garnet

The gravel on the bars of the Fraser River from Hope to Cache Creek provides the collector with jade, agate, jasper and garnets. There are many bars open to the public, but you must be aware of the many First Nations reserves and get permission. A good place to start checking for the gem stones is at the gravel bars right at the town of Hope. These bars are easily accessed and offer a wide variety of cutting material. Try the pullover at the weigh station on Highway 7 and walk across the tracks to the large bar. The mouths of the many creeks are also good, and you should take along a gold pan. Other good locations are at Emory Creek and at the town of Yale. This town has lots of history, and you can see some of that at the museum where they will also sell you a gold pan and other supplies you may have forgotten at home. As you head up river, the bars around Boston Bar and Alexandra are also worth a look. Each year the high water will turn over new material, so any place where you can get down to the river is worth a look. Check your copy of the B.C. recreational atlas or check with authorities if you are not sure of a bar's status.

The Fraser River at Boston Bar.

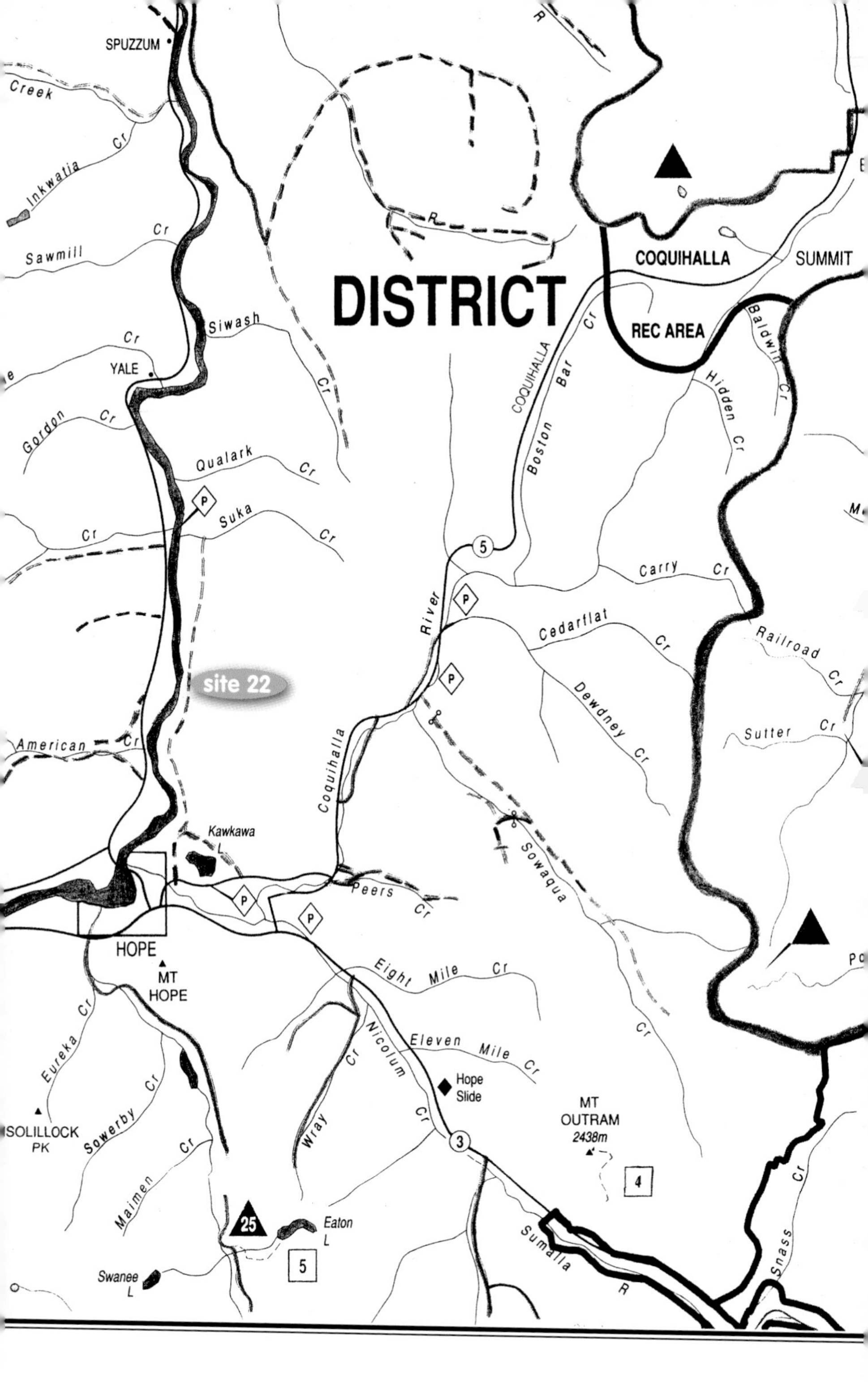

SPUZZUM
Creek
Inkwatia
Cr
Sawmill
Cr
Siwash
Cr
YALE
Gordon
Cr
Qualark
Cr
Suka
Cr
Cr
American
Cr
site 22
Kawkawa
L
HOPE
MT
HOPE
Eureka
Cr
SOLILLOCK
PK
Sowerby
Cr
Maimen
Cr
Swanee
L
25
Eaton
L
5
Wray
Cr
Nicolum
Cr
3
Eight Mile Cr
Eleven Mile Cr
Hope
Slide
MT
OUTRAM
2438m
4
Sumallo
R
Peers
Cr
Coquihalla
River
5
Sowaqua
Cr
Dewdney
Cr
Cedarflat
Cr
Carry
Cr
COQUIHALLA
Boston
Bar
Cr
DISTRICT
R
COQUIHALLA
REC AREA
SUMMIT
Baldwin
Cr
Hidden
Cr
Railroad
Cr
Sutter
Cr
Snass
Cr
R

JADE RESERVE

The following are the areas set aside by the BC government for public collecting of jade.

THAT, for many years the mineral nephrite which is more commonly known as jade has been found and recovered from the bed of the Fraser River where it flows between the towns of Lillooet and Hope.

And that the search for and recovery of jade from that area has been enjoyed for a number of years by the general public at large and in particular those persons who practice rockhounding, as a hobby.

And that it is desirable that jade in the area referred to herein be set aside for the benefit and enjoyment of the general public.

And to recommend that pursuant to subsection 5 of section 12 of the mineral act, and section 11 of the placer-mining act, and all other powers thereunto enabling, no person shall locate or record for the purpose of acquiring tittle to jade upon any of the lands described as follows:

All that part of the bed of the Fraser below mean high water mark between the centre lines of the Fraser River Bridge at Hope and the Lillooet Suspension Bridge.

And to further recommend that upon expiry of any placer-mining lease or mineral claim which lies within the above described area the lands covered by any such leasehold or mineral claim shall become and are hereby constituted a reservation under the terms of this order.

And to further recommend that it shall be lawful for any person to search for and recover jade from the area hereby reserved for his sole use and pleasure without the need for acquiring a Free Miners Certificate.

Dated this 25th day of January, 1968
Signed,
D. L. Brothers, Minister of Mines.
W.A.C. Bennet. Presiding member of the executive council.

JADE RESERVE BARS

- **The Yale group of bars** — below the town of Yale.
- **Alexander Bar** — east bank, 25.6 km north of Yale.
- **Anderson Bar** — east bank, 6.4 km south of Boston Bar.
- **North Bend Bar** — west bank, 4.8 km north of North Bend.
- **Kanaka Bar** — east bank, 27 km north of Boston Bar.
- **Van Winkle Bar** — west bank, 1.6 km north of Lytton.
- **Stein Bar** — west bank, 9.6 km north of Lytton.
- **17 mile Bar** — east bank, 27.2 north of Lytton.
- **28 mile Bar** — east bank, 44.8 north of Lytton.
- **Long Bar** — east bank, 4.8 km south of Lillooet.
- **Seton Creek Bar** — west bank, Lillooet.
- **Lillooet Bar** — west bank, Lillooet.

The Fraser River just north of Yale.

site 23 AINSLIE ROAD NORTH

▸ map page 43

agate

From Boston Bar, take Highway 1 for 11 km and turn right on Ainslie Road. Be careful, as there may be active logging. Keep to the left and proceed 29 km where the road to the right is marked 400. This road leads to Zatwatski Mountain, where there are supposed to be thunder eggs, but I have not found them yet. If you go past the road marked 400, and turn right on the next road, you will be on the road to the agates. The first agate location is at the 34-km marker. Here you will find agate in nodules, with some of them being black. The next location on this road is at the 36-km marker. In the road cuts and in the hills, you will find many agates with some of them larger than your fist. The main color we found here was a pale blue. The best location we have found, so far, is near the top of the hill, approximately at the 40-km marker area. The digging is easy and the rewards are very good. There are many rustic campsites with water near by. You are in the wild, so be sure to watch for bears as there have been many sightings in the area. This area is over 5,000 feet altitude, so collecting can only be done in the middle of summer.

Keep to the left and follow the forest service road.

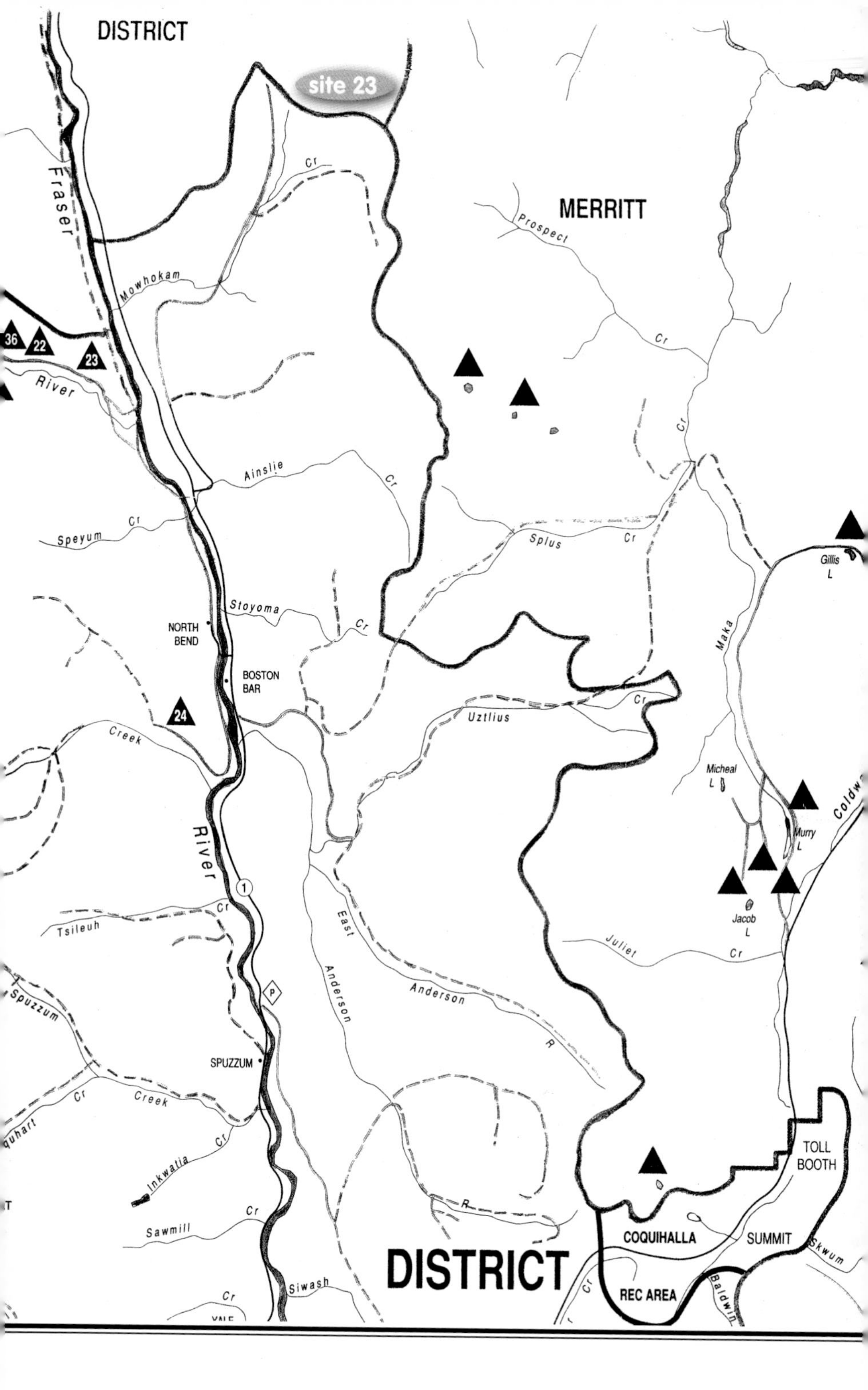

DISTRICT
site 23
MERRITT
Fraser
Prospect
Cr
Mowhokam
36
22
23
River
Ainslie
Cr
Speyum
Cr
Splus
Cr
Gillis
L
Stoyoma
Cr
NORTH
BEND
BOSTON
BAR
Maka
24
Creek
Uztlius
Cr
Micheal
L
Coldw
Murry
L
River
1
Jacob
L
Cr
Tsileuh
East
Juliet
Cr
Anderson
Spuzzum
Anderson
R
SPUZZUM
Cr
Creek
Cr
Inkwatia
TOLL
BOOTH
Cr
Sawmill
R
COQUIHALLA
SUMMIT
Skwum
DISTRICT
Cr
REC AREA
Baldwin
Siwash
Cr

▸ map page 45

soapstone

To get to the soapstone, take Highway 1 north from Hope to Boston Bar. Cross the Fraser River at Boston Bar and go through the small town of North Bend. Follow this road to Nahatlatch Valley. At 14.2 km from the river crossing, the soapstone is on the right side of the road. This location is on the road fill, so be sure to keep the road and ditches clear of debris. To find soapstone and serpentine in place, continue past the bridge, and turn right at the 14-km marker (.6 km). Follow this road for 1.2 km where there is a pile of serpentine beside the tracks. Park here, and walk less than half a kilometre to the left. The soapstone is in the large outcrop beside the tracks. In the Nahatlatch Lake area, there are many beautiful lakefront campsites. If you stay on the main road you will see the campsites all along the lakes. You must pack out your garbage. The snow can arrive here any time after late September so be sure to check before you go.

The bridge over the Nahatlatch River.

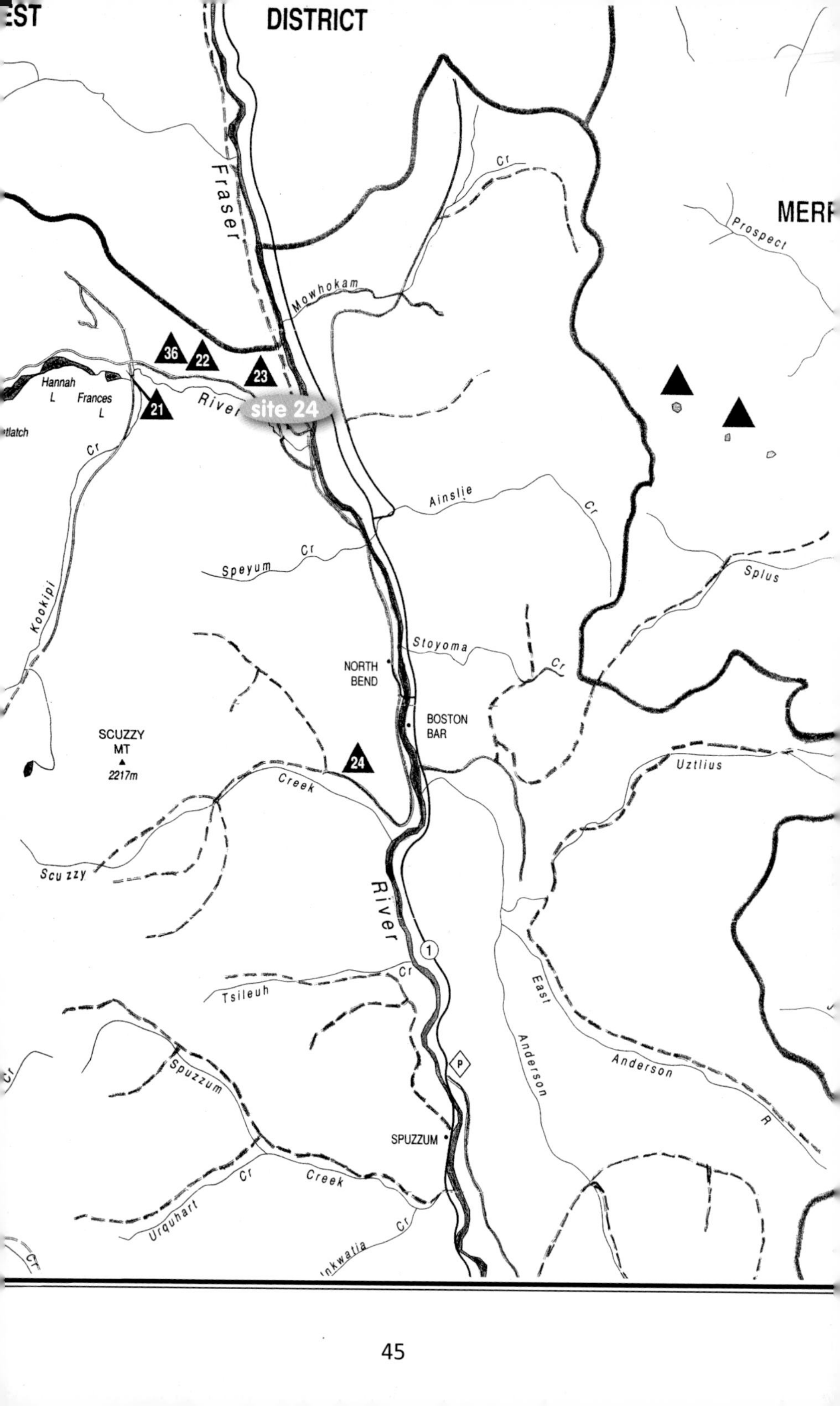
EST
DISTRICT
MERR
Fraser
River
Mowhokam
Cr
Prospect
Hannah
L
Frances
L
River
site 24
tlatch
Cr
36
22
23
21
Ainslie
Cr
Speyum
Cr
Splus
Kookipi
Stoyoma
Cr
NORTH
BEND
BOSTON
BAR
SCUZZY
MT
2217m
24
Creek
Uztlius
Scuzzy
River
1
Cr
Tsileuh
East
Anderson
Anderson
R
Spuzzum
P
SPUZZUM
Cr
Creek
Urquhart
Cr
Inkwatia

LOCAL CREEKS AND BARS

placer gold

Deposits of heavy gold were worked along the Fraser River from Hope to Lytton. Each year, the high water turns over some new material. Early spring is the best time to check the rocky bars. Dig under large boulders and around the inside curves of the river. When you pan for the gold, you will get garnets in your pan. Most of these are small but have very good clarity and color. Some major bars were Hills Bar at Yale and at Boston Bar. In the town of Yale the museum will give you a great deal of information on the area and the mining done here. Yale was at one time the biggest city north of San Francisco. It now has a population of well under 1,000. If you stop at the campsite at Emory Creek, they have a good write-up on the reclaiming that the Chinese did after the other miners left. Here you will see the old pits left by these workers. This campsite is a pay site. If you take the logging roads in this area there are many good rustic campsites, but watch out for logging trucks and leave the roads clear. There were some reports of gold on the Thompson River, but the Fraser River has proven to get results. Check with authorities about reserves and claims if you have any doubts about the bar you want to work.

site 25 SHAW SPRINGS SOUTH

▸ map page 49

smoky agate, plume agate

Head north out of Hope on Highway 1. Shaw Springs is 125 km north of Hope. Just 1.6 km south of Shaw Springs at the cement culvert, pull off the road and climb the railroad track bank. One good spot to get agate nodules is in the rock bluff across the tracks and to the right of the culvert. It is down the tracks a few hundred metres. As you walk the tracks, you will see the rock bluff that has indications of an underground spring near it. Most of the agates at this location will be clear or will have a blue tinge to them. The outside covering is a dull green. There are many agates in this area and you should check the bluffs around you. I have heard that there are snakes in this area, but in the trips I've made, I have not seen any. If you see a rattler, give it plenty of room and try another spot, as there are many good agate locations in this area.

A Fraser River bar near Yale.

The culvert at the Shaw Springs collecting site.

site 26 SANTA CLAUS ROCK

▸ map page 49

agate, opal, calcite

From the town of Lytton on Highway 1, head east 29.5 km to the forest service road heading into the hills on your right. The road will take you to Santa Claus Rock. The collecting area is 6.2 km in. The agates can be found on Santa Claus rock itself and across a small valley in the low cliffs. These cliffs are above the road and on your left. Just before you get to the big rock face you will keep left and the road crosses to the left-hand side of the small valley. Agates, opal and calcite are inside the red skin-covered rock. Some pink amethyst has also been found here. There is more agate in the area so check any small signs you may see on the roads you travel. [Note: A four-wheel-drive vehicle is needed to get to this site.]

Santa Claus rock.

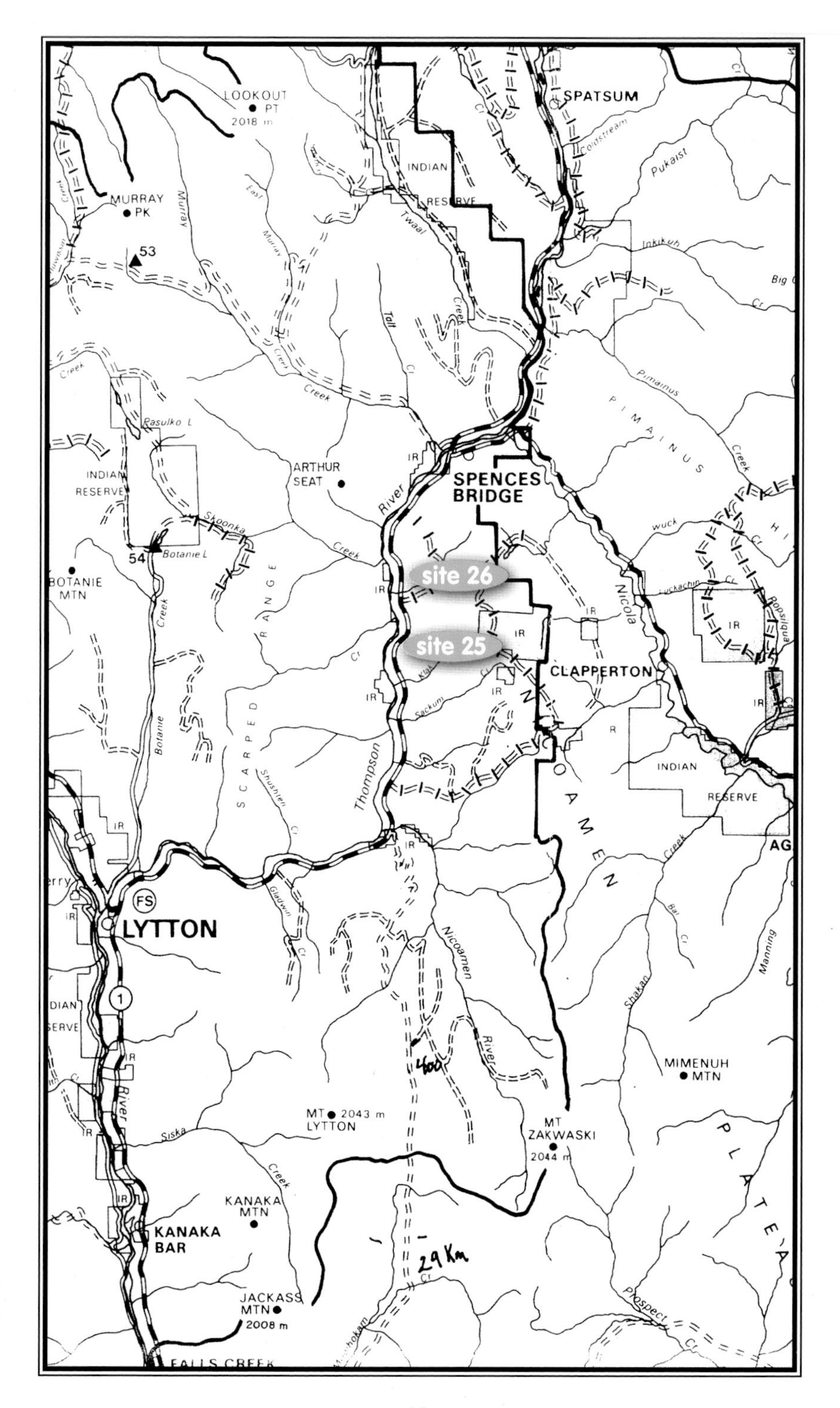
LOOKOUT PT 2018 m
SPATSUM
MURRAY PK
53
INDIAN RESERVE
ARTHUR SEAT
SPENCES BRIDGE
PIMAINUS
site 26
site 25
CLAPPERTON
54
BOTANIE MTN
SCARPED RANGE
NICOAMEN
LYTTON
INDIAN RESERVE
MIMENUH MTN
MT 2043 m LYTTON
MT ZAKWASKI 2044 m
PLATEAU
KANAKA MTN
KANAKA BAR
JACKASS MTN 2008 m
FALLS CREEK
400
29 Km

site 27 VENABLES VALLEY

▸ map page 51

agate, petrified wood, geodes

From Venables Valley to Highway 12 (west of Spences Bridge to Cache Creek), the back roads provide many locations for gem and mineral samples. One of the best choices is the southern entrance to Hat Creek Road on the west side of Highway 1. This road goes from 21.4 km north of Spences Bridge through to Highway 12. Parts of the road are steep, so be careful and watch for other traffic. After you get into the hills, where ever there is rock bluff or creek, check for agates and petrified wood. The spots are mainly on the right-hand side of the road. If the spot you stop at has no signs of the gems, proceed to the next good looking spot. There are many nice camping spots along the way. For agate, be sure to check all likely locations, especially near large rock outcrops. One location for agate and jasper-agate is in Medicine Creek and in the outcrop above the creek. Winter can be a problem, as can a heavy rainfall, so be sure to check weather reports before you leave.

Just before the south turnoff of Hat Creek Road.

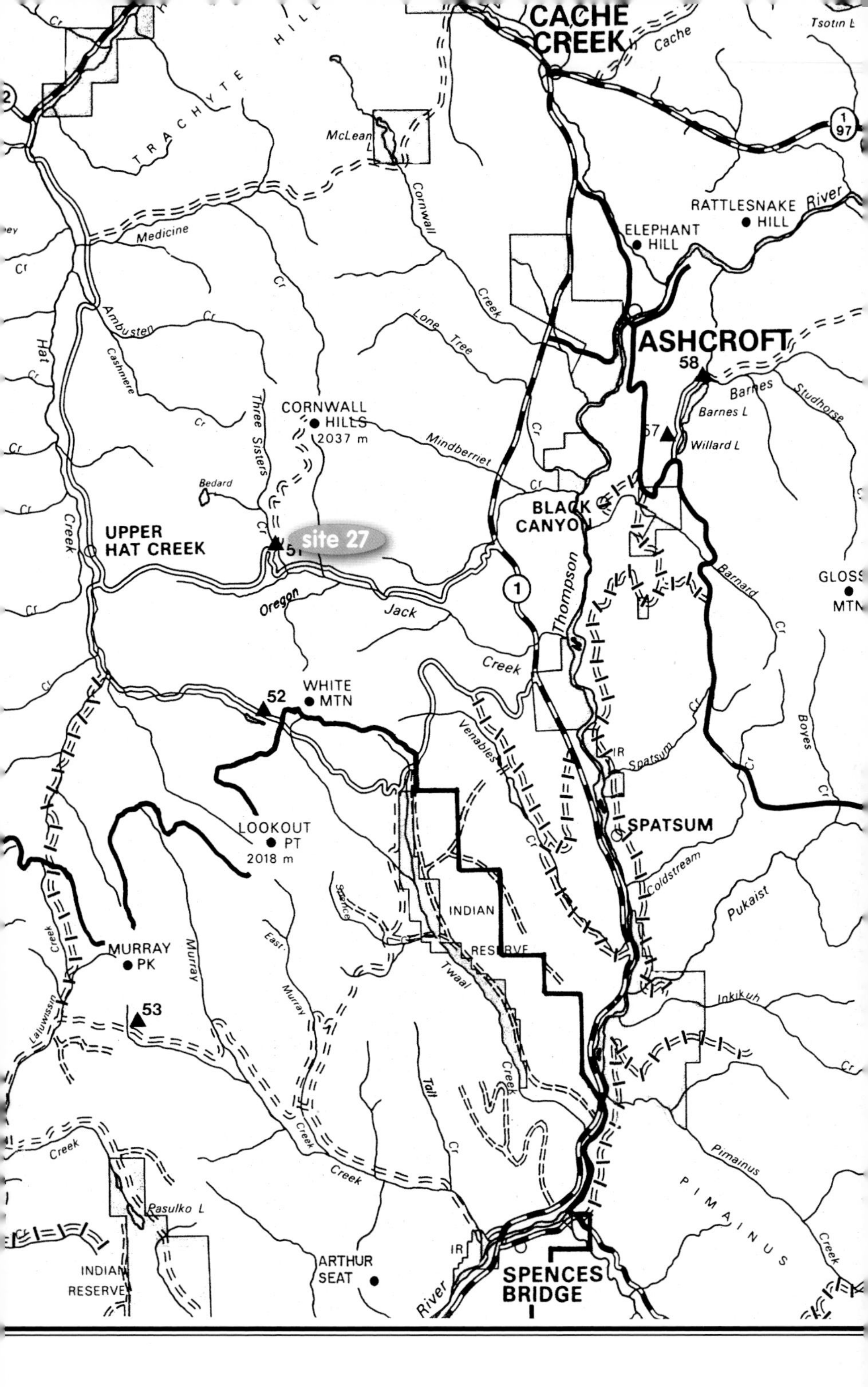
CACHE CREEK
Cache
ASHCROFT
58
RATTLESNAKE HILL
ELEPHANT HILL
Rattlesnake River
TRACHYTE HILLS
McLean L
Cornwall Creek
Medicine
Ambusten Cr
Cashmere Cr
Hat Creek
Lone Tree
Three Sisters Cr
CORNWALL HILLS 2037 m
Mindberriet Cr
Bedard
UPPER HAT CREEK
51
site 27
BLACK CANYON
Barnes L
Willard L
57
Barnes
Studhorse
Thompson
Oregon Jack Creek
Barnard Cr
GLOSS MTN
WHITE MTN
52
Venables Cr
Spatsum Cr
SPATSUM
LOOKOUT PT 2018 m
Coldstream
Pukaist
INDIAN RESERVE
Twaal Creek
Spence
East Murray
Murray Creek
MURRAY PK
53
Laluwissin Creek
Inkikuh Cr
Tait Cr
Pimainus
PIMAINUS Creek
Pasulko L
INDIAN RESERVE
ARTHUR SEAT
IR
River
SPENCES BRIDGE
Boyes Cr
Tsotin L
1
97

site 28 BACK VALLEY ROAD ▸ map page 53

agate

From Cache Creek, head east on Highway 1 for 2.7 km to Back Valley Road. Go north for 3.8 km and park on the right, just after the bridge. Cross back over the bridge and walk east for 600 metres. Walk up the large draw about 600 metres to find agate in the rock outcroppings. The main agate that is here is tube agate, but there is some nice moss agate as well as some zeolite. There will be other locations near here. Check the creek beds and the bottom of the draws for any hints of where to look for agate. Also, there is a pipeline that appears near the road from time to time. Check the pipeline for agates where there is a cut in the banks. There are forest service campsites on some of the lakes in the area that are great for camping.

Back Valley Road turnoff.

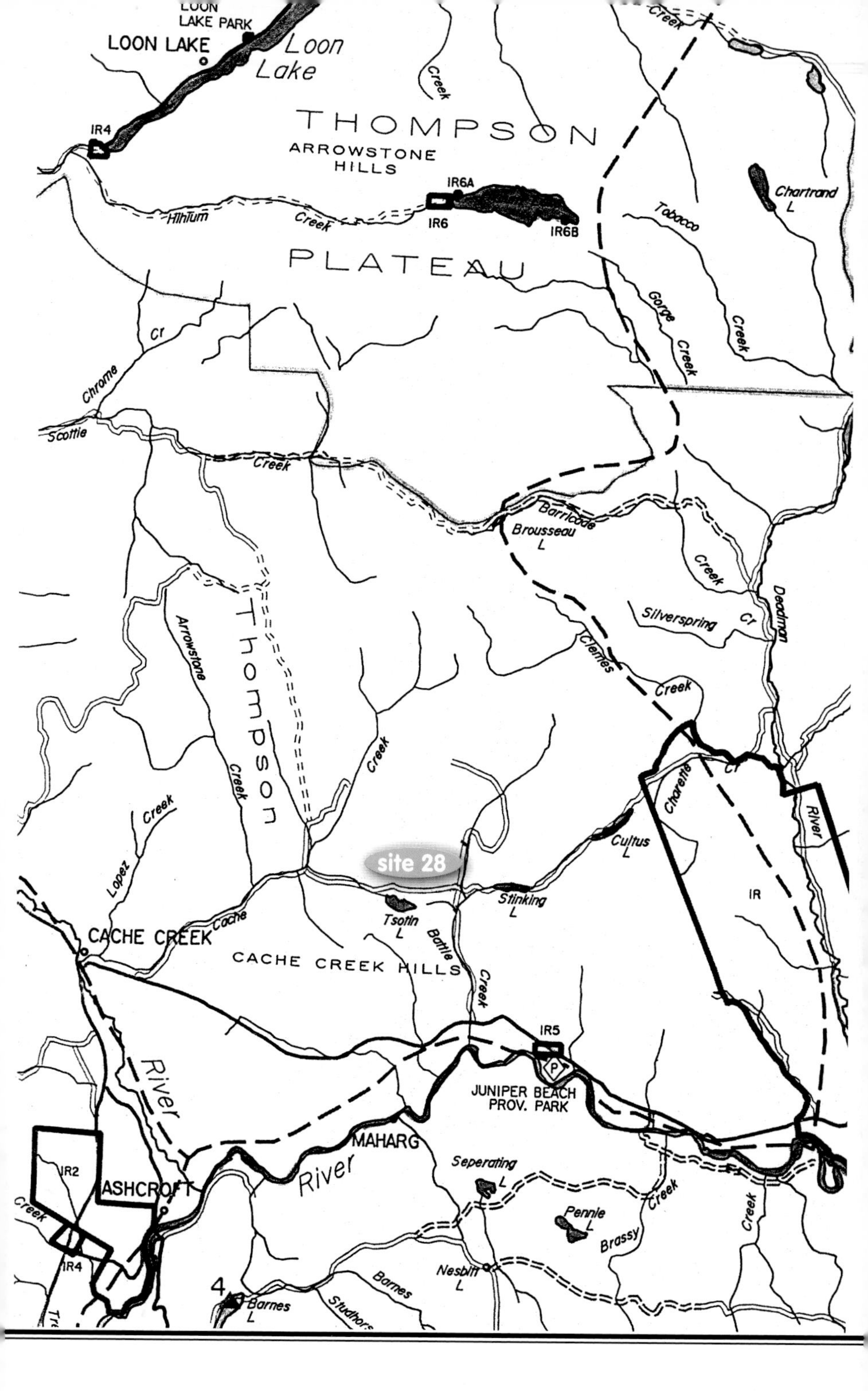
LOON LAKE PARK
LOON LAKE
Loon Lake
IR4
THOMPSON
ARROWSTONE HILLS
IR6A
IR6
IR6B
Hihium
Creek
PLATEAU
Chartrand L
Tobacco
Gorge Creek
Creek
Chrome Cr
Scottie Creek
Barricade
Brousseau L
Creek
Silverspring Cr
Deadman
Clemes Creek
Arrowstone Creek
Thompson
Creek
Charette Cr
River
Cultus L
site 28
IR
Lopez Creek
Stinking L
Tsotin L
Cache
CACHE CREEK
CACHE CREEK HILLS
Battle Creek
IR5
JUNIPER BEACH PROV. PARK
River
MAHARG
River
Seperating L
IR2
ASHCROFT
Pennie L
Creek
Brassy
Creek
IR4
Barnes
Nesbitt L
Barnes L
Studhorse

site 29 DEADMAN CREEK 1

▸ map page 55

petrified wood, agate, crystal-lined vugs

This is a great place to collect rock samples and get some good pictures of beautiful rock formations. On this road you will come across hoo-doos, large cliffs, and many other interesting formations. Heading east from Cache Creek on Highway 1 travel to Deadman Creek Road, 7.4 km west of Savona. Head north, and at 15 km you will cross over Deadman Creek and will be traveling alongside the pipeline. When you are at the 20-km marker, you will see two main shutoffs for the pipeline, and it is here that you will turn on the pipeline access road. This is forestry road 552 (McLeod Road). At 1.2 km in, you will find agate used as road fill and in the road cut. The source of the agate is the steep cliffs directly above you. Continue on this road until you are closest to the cliffs. The petrified wood is at the base of the large bluffs. Some great agate nodules are found on the pipeline just around the 2-km area on a road heading off to your right.

The McLeod Road turnoff.

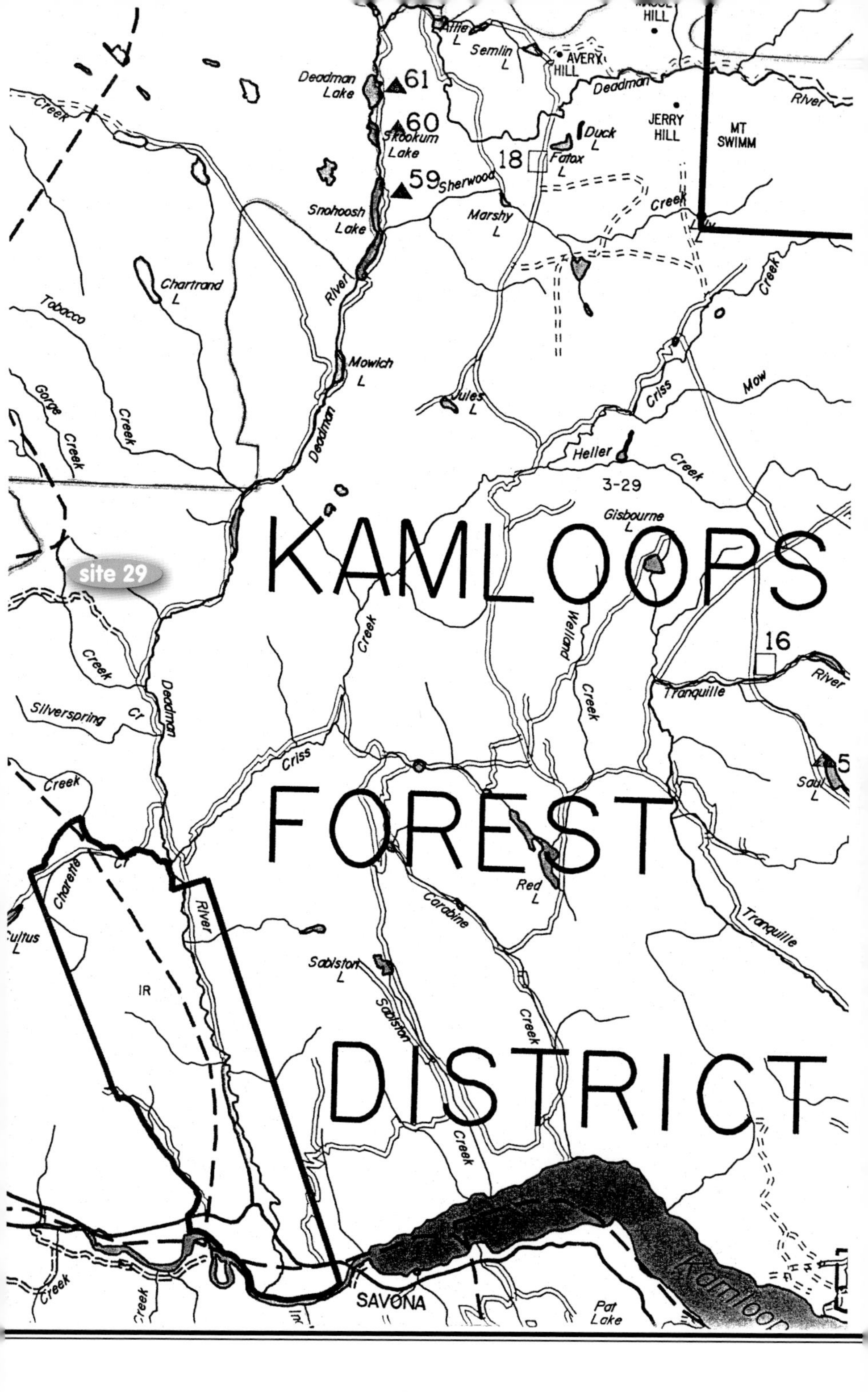
KAMLOOPS
FOREST
DISTRICT
site 29
Deadman Lake
Skookum Lake
Snohoosh Lake
Allie L
Semlin L
AVERY HILL
JERRY HILL
MT SWIMM
Deadman River
Duck L
Fatox L
Sherwood Creek
Marshy L
Chartrand L
Tobacco
Gorge Creek
Creek
Mowich L
Jules L
Criss
Mow
Heller
Creek
3-29
Gisbourne L
Welland Creek
16
Tranquille River
Saul L
Silverspring Cr
Criss
Charette Cr
River
IR
Cultus L
Carabine Creek
Red L
Sabiston L
Sabiston Creek
Tranquille
SAVONA
Pat Lake
Kamloo
61
60
59
18

site 30 DEADMAN CREEK 2 ▸ map page 57

agate, druzy quartz

This is a great place to collect rock samples and get some good pictures of beautiful rock formations. On this road you will come across hoo-doos, large cliffs, and many other interesting formations. Heading east from Cache Creek on Highway 1 travel to Deadman Creek Road, 7.4 km west of Savona. Head north, and at 15 km you will cross over Deadman Creek and will be traveling alongside the pipeline. When you are at the 20-km marker, you will see two main shutoffs for the pipeline, and it is here that you will turn on the pipeline access road. This is forestry road 552 (McLeod Road). From Deadman Creek Road you will travel on this road for about 5 km and here you will cross the pipeline again. The pipeline intersects the road, and you will see a steep hill following the pipeline. Go up the hill and the agates and crystals are on the left, .5 km from the intersection.

The agates are near the top of the road cut.

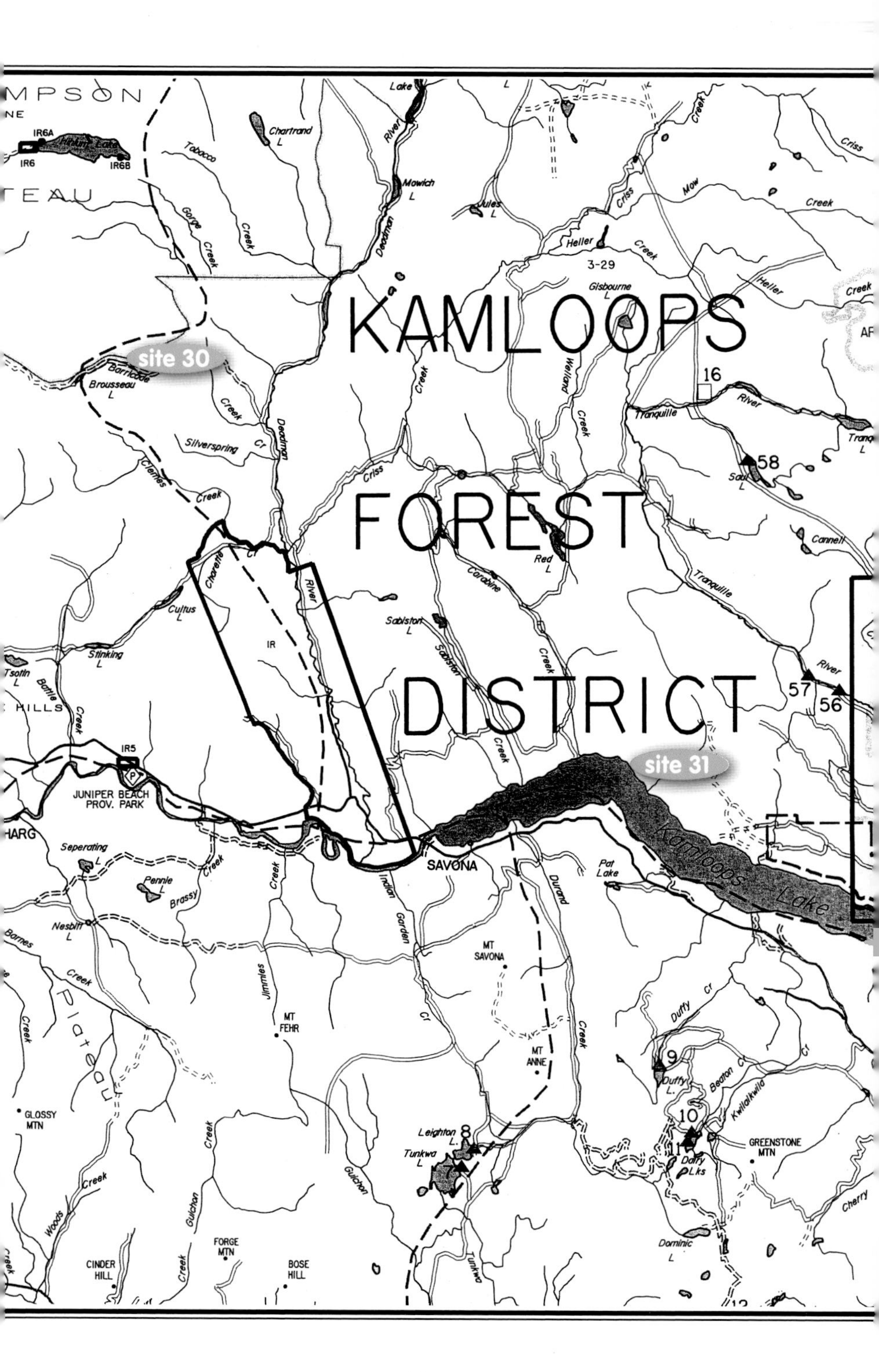
KAMLOOPS
FOREST
DISTRICT
site 30
site 31
SAVONA
Kamloops Lake
JUNIPER BEACH PROV. PARK
Chartrand L
Mowich L
Jules L
Heller L
Gisbourne L
Brousseau L
Cultus L
Stinking L
Sabiston L
Red L
Saul L
Cannell L
Pat Lake
Seperating L
Pennie L
Nesbitt L
Leighton L.
Tunkwa L
Duffy L.
Duffy Lks
Dominic L
MT SAVONA
MT FEHR
MT ANNE
GLOSSY MTN
GREENSTONE MTN
FORGE MTN
CINDER HILL
BOSE HILL
Deadman River
Tranquille River
58
57
56
16
3-29
9
10
8
7

site 31 KAMLOOPS LAKE ▸ map page 57

agate, calcite, fossils

Kamloops Lake provides rockhounds with many good collecting areas. Road access is the main problem when collecting here. You can follow the railroad tracks from the main highway across from Savona. Also in Savona, there are government campsites with water, washrooms and tables. If you walk along the beaches on the north side of the lake, when the water is low, you can find arrow heads and other artifacts. There have been other finds including fish fossils, agate and other collectables. Locations for these rocks are hard to describe because the finds are not in any one location and are found in ones or twos. While at the government camps, the beaches of Kamloops Lake are a good location for agate and opal. The opal is scarce, but can be found at the beaches right in the town of Savona.

Overlooking Kamloops Lake and surrounding area.

site 32 HAYWOOD FARMER ROAD

▸ map page 61

agate, jasper

From Cache Creek continue east on Highway 1 toward Kamloops. Just east of Savona, go south on Tunkwa Lake Road. This road winds up the hill away from Kamloops Lake, providing good opportunities for pictures. At 3.6 km turn west on Haywood Farmer Road and go for 5.8 km where you will come to a cross road. Turn left and travel for 3.1 km where there is a road cutting back on your left. If you drive down this road for half a kilometre, you will see agate in the road cut. The host rock in this area will break up with a small rock hammer making the agates easy to get. On the main road in this area, there is a pipeline you can follow and agate can be found at different locations along it. There are many good camping spots in this area, but be sure to keep your food locked up. The last time I was in the region there was a lot of bear sign. The roads in this area are not good after a rain, as they become very mucky.

Agates in the roadcut.

The trail down to the opal is to the right.

site 33 SAVONA TOWERS

▸ map page 61

opal

From Cache Creek continue east on Highway 1 toward Kamloops. Just east of Savona, go south on Tunkwa Lake Road. This road winds up the hill away from Kamloops Lake. When you are 12 km from the highway, go west to the towers. The towers are at the end of the road, at the top of the hill. Follow the foot trail west down the steep grade for half of a kilometre. The trail is located to the right of the hang-glider jump platform and is quite steep. Where the trail diminishes, you are on the basalt. This is where you will find opal. Also, in the middle of the draw, and on the far side, more opal will be found. These locations have less traffic. The roads in are good, but do not travel them after a heavy rain. There are lots of good campsites with creeks nearby, but you may want to bring water with you, depending on the time of the year. During the summer some creeks might be dry. There are campsites at Tunkwa Lake, with picnic tables and great fishing in the spring and early summer. There is a lot of agate in this area, so be sure to check any outcrops or small streams for signs of where to look (small agate chips).

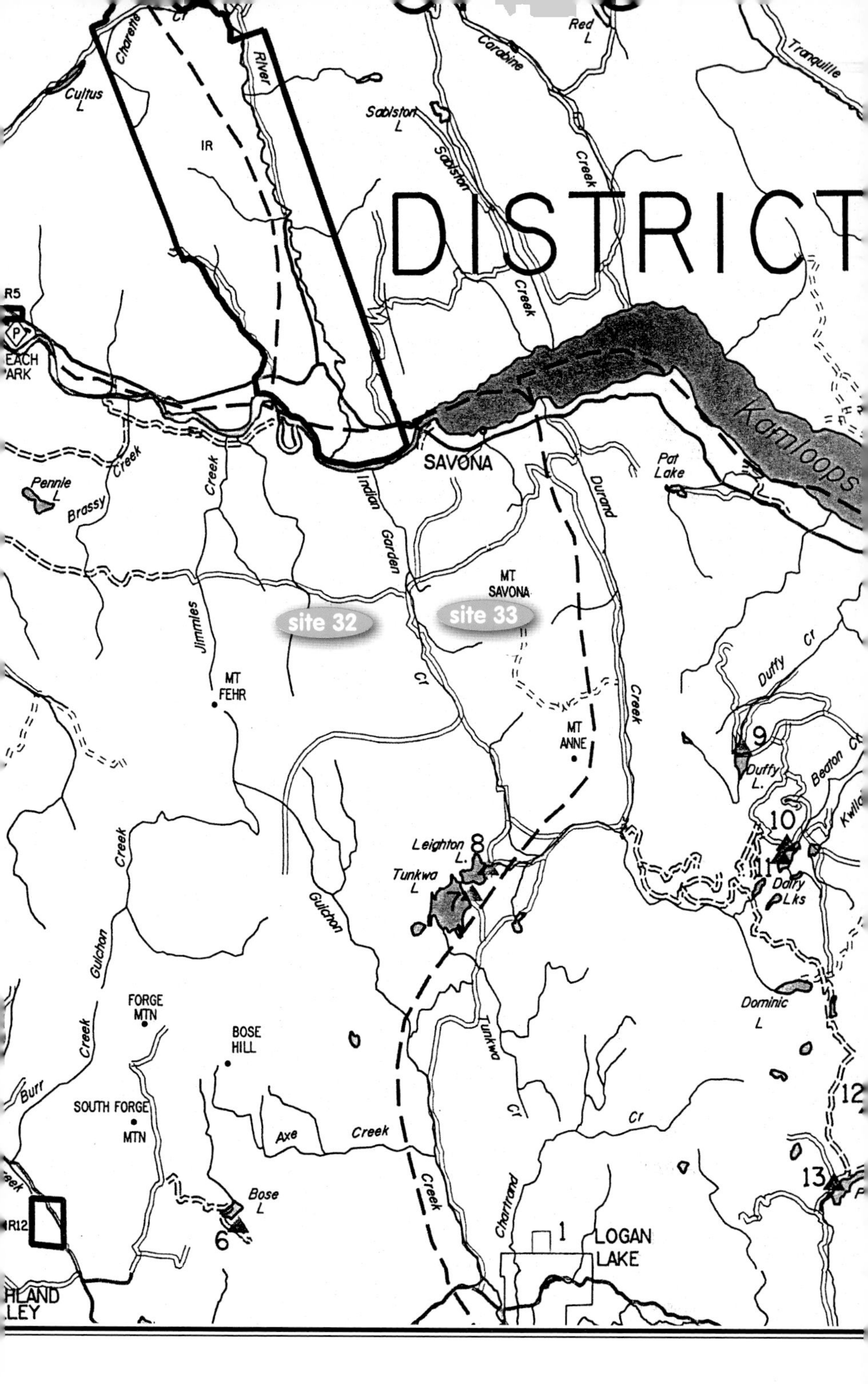
DISTRICT
Charette Cr
River
Red L
Carabine
Tranquille
Cultus L
Sabiston L
Sabiston
Creek
IR
R5
P
EACH ARK
Creek
Kamloops
SAVONA
Pat Lake
Pennie L
Brassy Creek
Creek
Indian Garden Cr
Durand Creek
MT SAVONA
site 32
site 33
Jimmies
MT FEHR
Duffy Cr
9
Duffy L.
Beaton Cr
MT ANNE
10
Leighton L.
8
Tunkwa L
7
Dairy Lks
Creek
Gulchon
Gulchon
Dominic L
FORGE MTN
Creek
BOSE HILL
Tunkwa Cr
Burr
SOUTH FORGE MTN
Axe Creek
Cr
12
Creek
Chartrand
13
Bose L
6
IR12
1
LOGAN LAKE
HLAND LEY

site 34 HIGHWAY 99

▸ map page 63

jasper, agate

From Cache Creek, head north on Highway 97 for 10.7 km to Highway 99, which goes to Pemberton. At 21.4 km in, the Hat Creek Road forks off to the left. Park there, and along the ridge to the north of the highway there is jasper and agate. They are good quality and take a good polish. They are found as float on the way up to the top of the ridge.

site 35 HAT CREEK

▸ map page 63

jasper, agate, petrified wood

Head north on Highway 97 for 10.7 km to Highway 99. At 21.4 km, turn left on Hat Creek Road. At 3.2 km there are some coal beds on the right. The petrified wood found here is not polishing material but makes for good displays. At 6 km from the highway, on the Hat Creek Road there is jasper and agate up the creek. This is Medicine Creek. Check up the forest service road, on the left, for more material. All along Hat Creek Road back to Highway 1 there are other opportunities for agate and jasper.

The north end of Hat Creek.

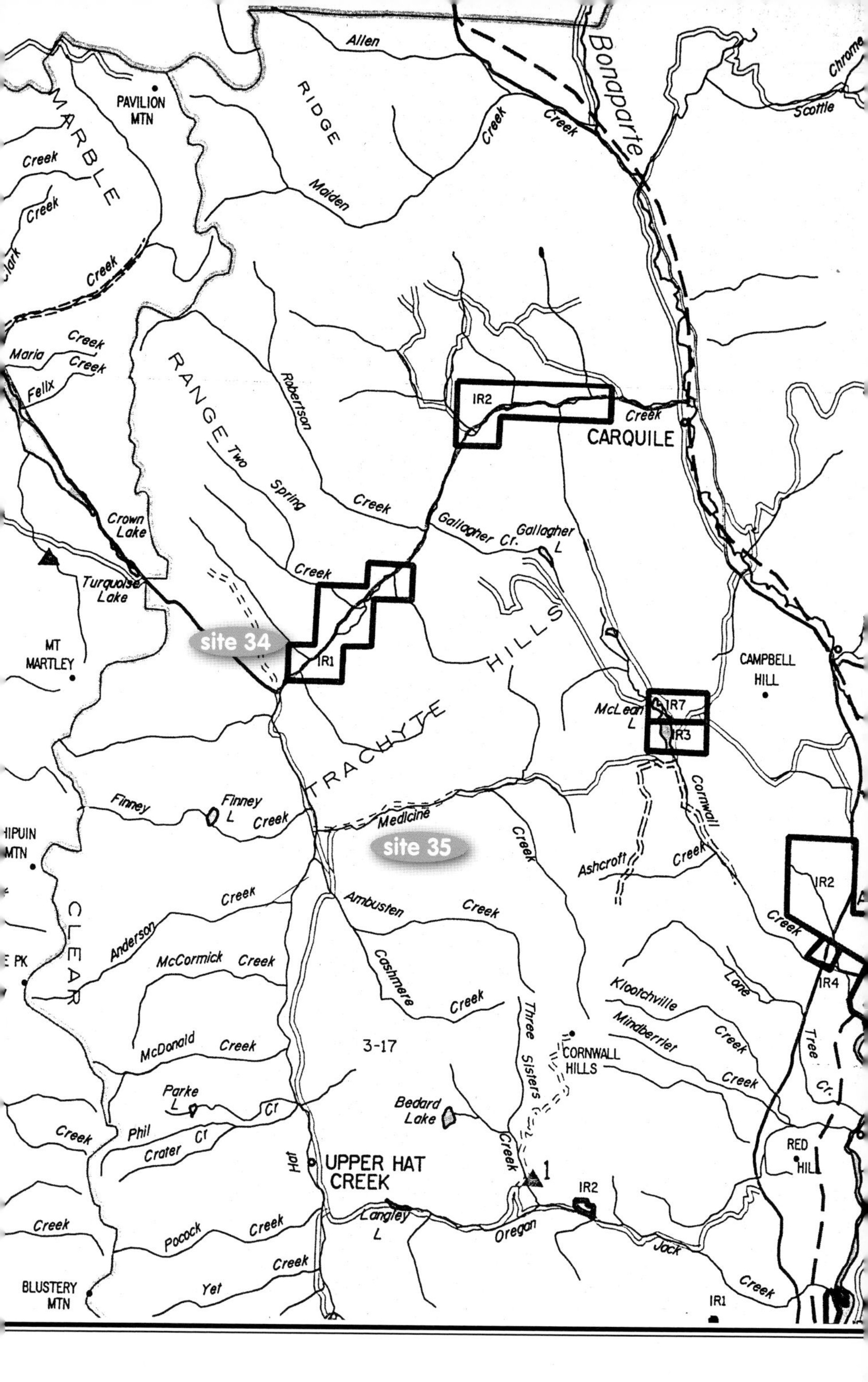
Allen
RIDGE
PAVILION MTN
MARBLE
Bonaparte
Creek
Chrome
Scottie
Maiden
Creek
Clark
Maria
Felix
RANGE
Robertson
Two
Spring
Creek
IR2
CARQUILE
Gallagher Cr.
Gallagher L
Crown Lake
Turquoise Lake
MT MARTLEY
site 34
IR1
HILLS
TRACHYTE
CAMPBELL HILL
McLean L
IR7
IR3
Cornwall
Finney
Finney L
Medicine
site 35
Ashcroft
Ambusten
Anderson
McCormick
CLEAR
Cashmere
Klootchville
Mindberriet
Lane
IR4
Tree Cr.
McDonald
3-17
Three Sisters
CORNWALL HILLS
Parke L
Phil
Crater
Bedard Lake
Hat
UPPER HAT CREEK
RED HILL
Langley L
Oregon
Jack
Pocock
Yet
BLUSTERY MTN

Just before the Scotty Creek turnoff.

site 36 SCOTTY CREEK

▸ map page 65

amethyst crystals, opal

From Cache Creek, travel Highway 97 north for 30 km to the Scotty Creek forest service road. Follow this road east for 7 km. At this point you will be making a sharp turn to your right. The crystals are in the small knob on the right-hand side of the road. Although the pieces are large and pretty, most of them will fade in sunlight. Included with the amethyst crystals, in some veins, is a common opal. There are other diggings in the area, so be sure to take a look around. Across from the location mentioned, there is a large field that is great for camping. This is free-range cattle country so keep your dog at bay if cattle are present. Just before the hill there is a great camping spot on the right.

site 37 BONAPARTE RIVER VALLEY

▸ map page 65

chalcedony, opal

Throughout the Bonaparte River Valley, where serpentine occurs, there will also be some chalcedony and opal.

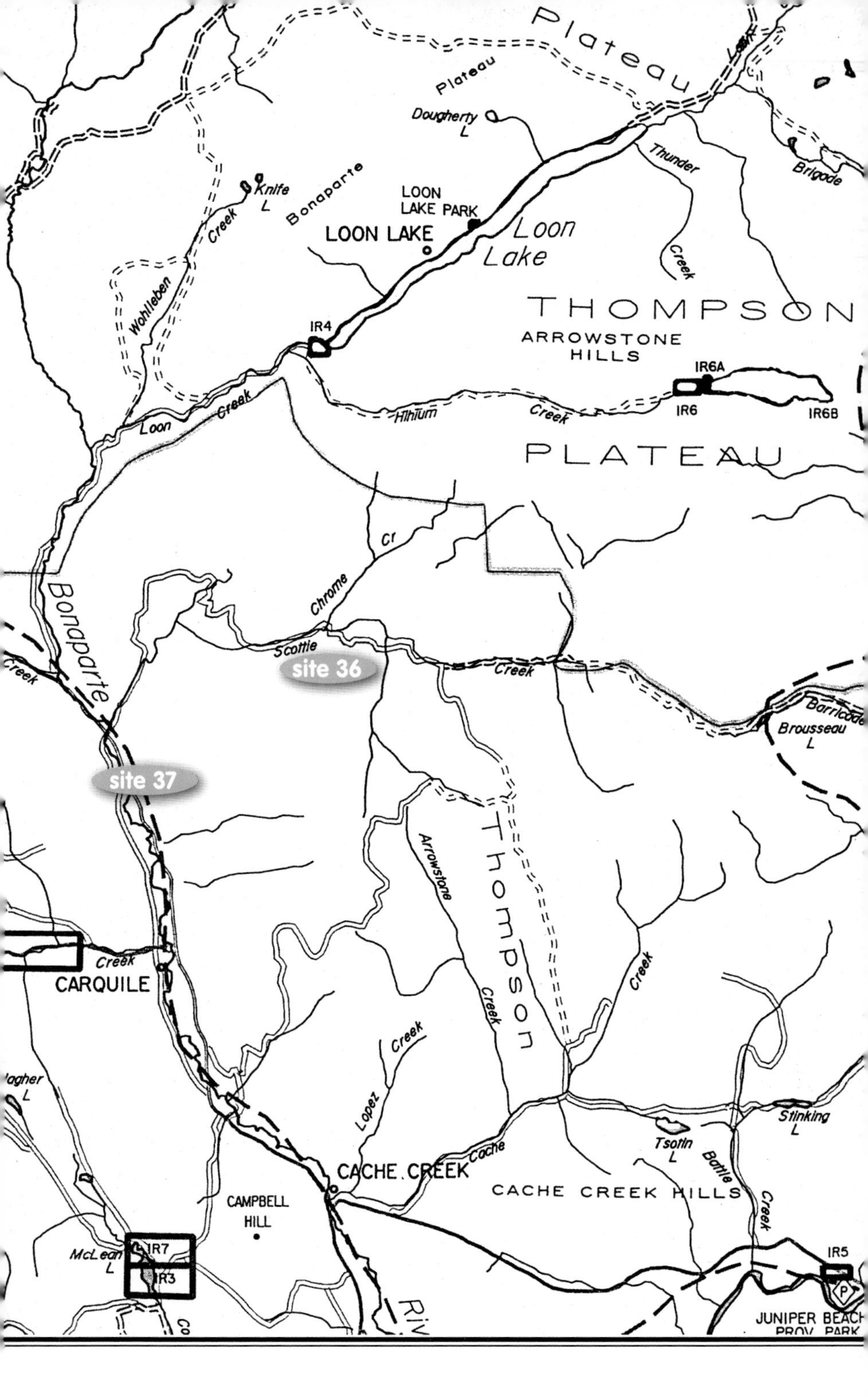
Plateau
Plateau
Dougherty L
Knife L
Bonaparte
Creek
LOON LAKE PARK
LOON LAKE
Loon Lake
Thunder
Brigade
Creek
Wohlleben
IR4
THOMPSON
ARROWSTONE HILLS
IR6A
IR6
IR6B
Loon
Creek
Hihium
Creek
PLATEAU
Cr
Chrome
Bonaparte
Creek
Scottie
Creek
site 36
Barricade
Brousseau L
site 37
Arrowstone
Thompson
Creek
Creek
Creek
CARQUILE
Creek
Lopez
Creek
Cache
CACHE CREEK
CACHE CREEK HILLS
Tsotin L
Battle
Stinking L
Creek
CAMPBELL HILL
McLean L
IR7
IR3
IR5
JUNIPER BEACH PROV. PARK

site 38 CLINTON

▸ map page 67

travertine, fossils

From Cache Creek, head north for 40 km on Highway 97 to Clinton. As you enter the town of Clinton, there is a road heading west. This is Pavilion Road. If you follow this road for 3.8 km, there is an old limestone quarry on the right-hand side of the road. The best material is on the railroad allowance. Most of this material is a golden color; it is very pretty and takes a nice polish. Be sure to keep the railroad clear of all debris. Also in this area, there are fossils of leaves and branches. A good place to start your search for fossils is in the railroad cuts. When you find a spot be sure to keep the tracks clear of the rocks you are digging. There are many good campsites in the area with wood and water. Also in the town of Clinton, there are fee campsites with tables, water and showers.

Pavillion Road turns to the left.

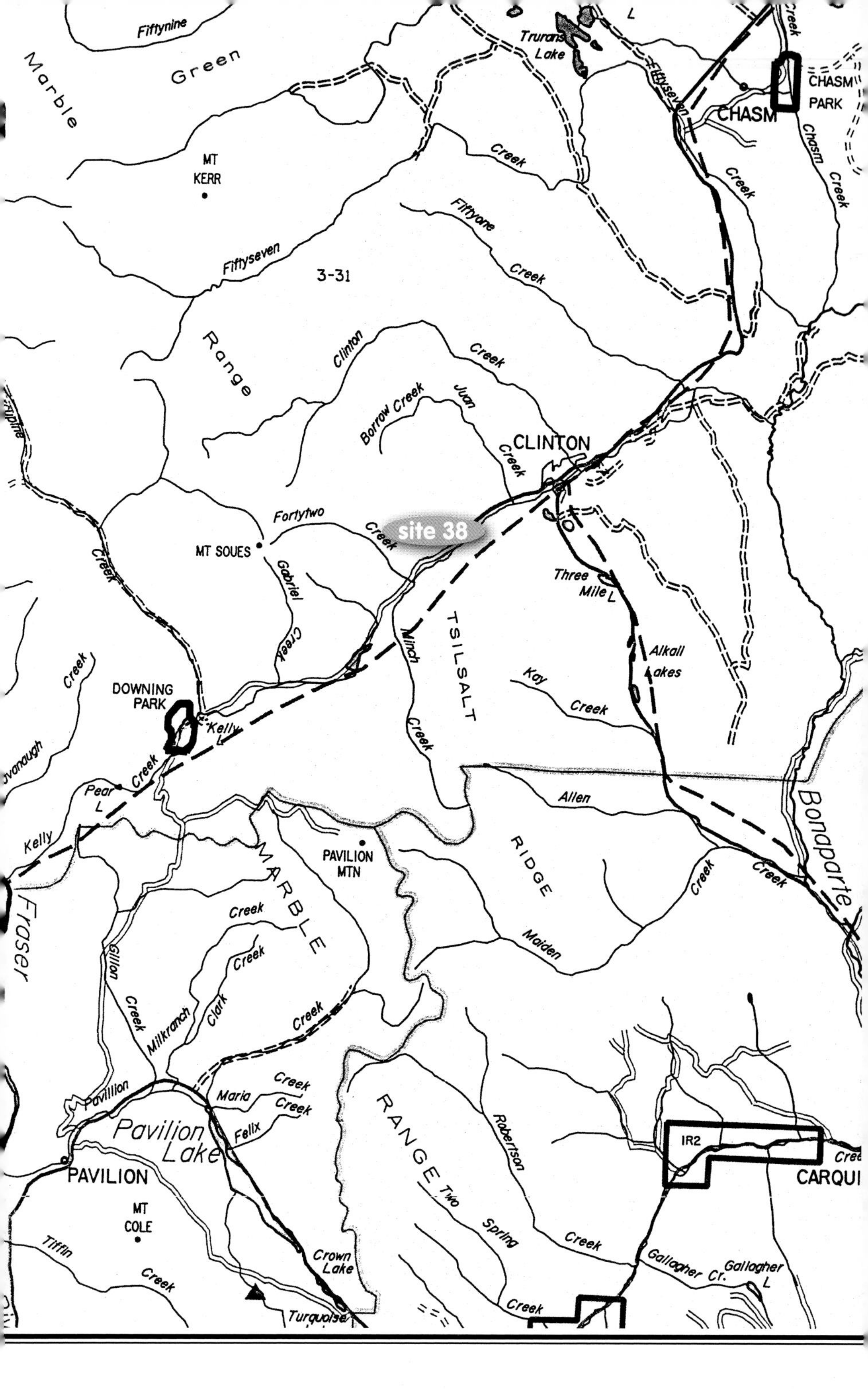
Fiftynine
Marble
Green
Trurans Lake
Fiftyseven
CHASM
CHASM PARK
Chasm
Creek
MT KERR
Fiftyone
Creek
Fiftyseven
3-31
Range
Clinton
Creek
Borrow Creek
Juan
CLINTON
Creek
Fortytwo
Creek
site 38
MT SOUES
Gabriel
Creek
Three Mile L
TSILSALT
Minch
Creek
Alkali Lakes
Kay
Creek
DOWNING PARK
Kelly L
Creek
Pear L
Kelly
Allen
Bonaparte
RIDGE
PAVILION MTN
MARBLE
Fraser
Creek
Creek
Maiden
Creek
Gillon
Creek
Clark
Milkranch
Creek
Pavillion
Maria
Creek
Felix
Pavilion Lake
PAVILION
RANGE
Robertson
IR2
CARQUI
MT COLE
Two
Spring
Creek
Tiffin
Creek
Crown Lake
Gallagher Cr.
Gallagher L
Creek
Turquoise

site 39 CHASM

opal

North of Cache Creek on Highway 97 is the town of Clinton. Staying on Highway 97 head north out of Clinton. At 17 km there is the Chasm Road turnoff. This is a great place to visit as the chasm seems to appear out of nowhere. Park at the lookout; the opal is at the head of the chasm. This is also a great place to check for more opal locations, as well as other material. If where you want to dig is still in Chasm Park, permission must be obtained. The forestry roads in the area will provide you with good camping sites, and the lakes and streams in the area are great for fishing. There are numerous forestry camps around, and most of them are at or near the lakes. A map of all the campsites in these areas can be obtained from the local forestry office.

Overlooking the chasm from Chasm Park.

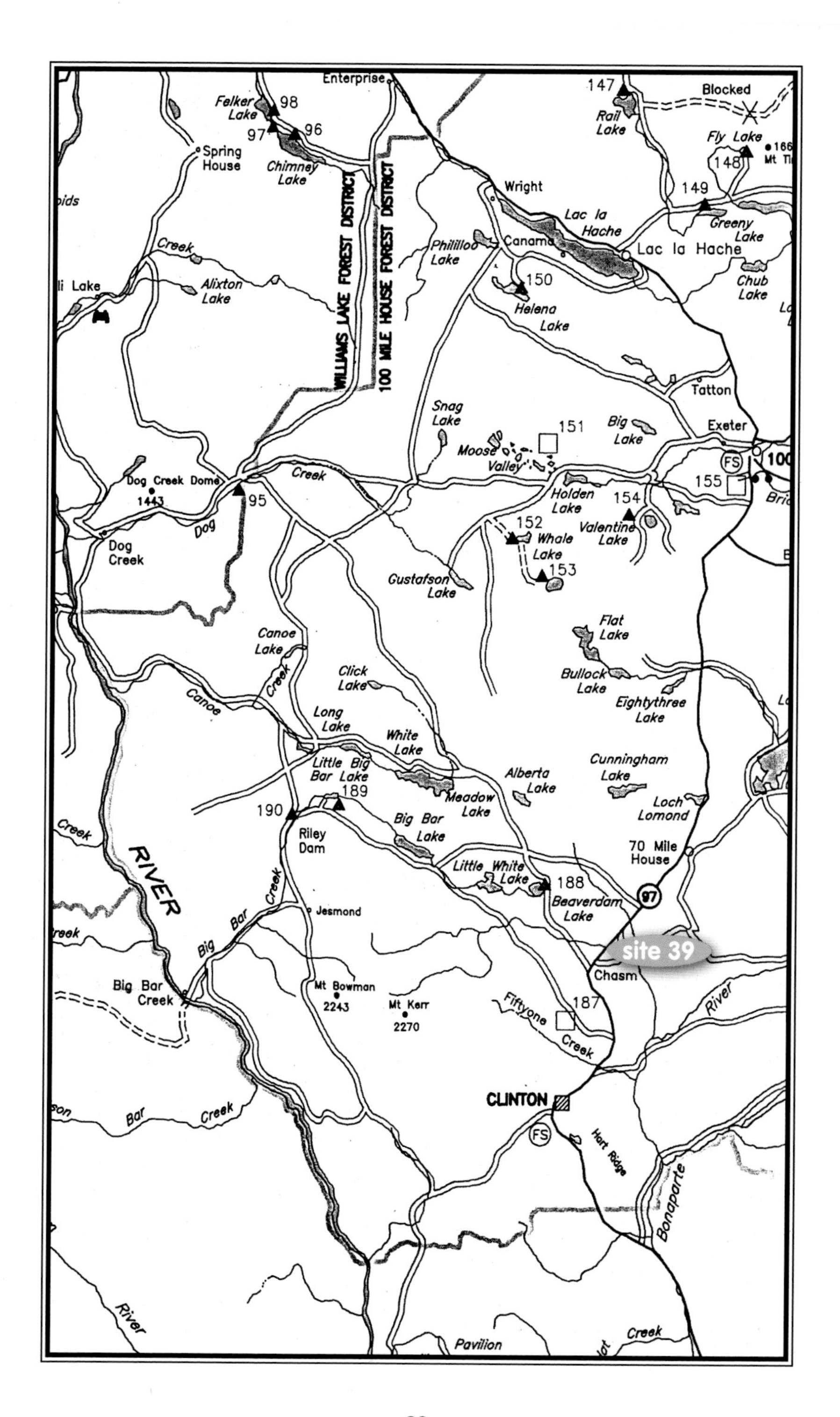

Enterprise
Felker Lake
98
97
96
Chimney Lake
Spring House
WILLIAMS LAKE FOREST DISTRICT
100 MILE HOUSE FOREST DISTRICT
147
Rail Lake
Blocked
Fly Lake
148
149
Wright
Lac la Hache
Lac la Hache
Greeny Lake
Chub Lake
Phililloo Lake
Canama
150
Helena Lake
Creek
Alixton Lake
Tatton
Snag Lake
151
Big Lake
Exeter
Moose Valley
155
Holden Lake
154
Valentine Lake
152
Whale Lake
153
Dog Creek Dome
1443
Creek
95
Dog
Dog Creek
Gustafson Lake
Flat Lake
Bullock Lake
Eightythree Lake
Canoe Lake
Creek
Canoe
Click Lake
Long Lake
White Lake
Little Big Bar Lake
189
190
Riley Dam
Alberta Lake
Cunningham Lake
Meadow Lake
Loch Lomond
Big Bar Lake
70 Mile House
Little White Lake
188
Beaverdam Lake
97
RIVER
Creek
Big Bar Creek
Jesmond
site 39
Chasm
Big Bar Creek
Mt Bowman
2243
Mt Kerr
2270
Fiftyone Creek
187
River
CLINTON
FS
Hart Ridge
Bonaparte
Bar Creek
River
Pavilion
Creek

site 40 BLACK DOME ▸ map page 71

thunder eggs, common opal

From Clinton head north on Highway 97, and at 17 km turn west on Meadow Lake Road. For this location you will need extra gas, and it is smart to go in the summer when the roads are at their best. Travel 95 km and cross the Fraser River at Dog Creek, then turn right on Black Dome Road (just after Brown Lake). Proceed 18.2 km and turn left on the road. Follow this road for about 8 km. Here you will turn left on Perlite Road following it 7.4 km to the end. Here there is a pit containing opal, thunder eggs and perlite. A lot of this terrain was part of the Gang Ranch, which was, once, the biggest ranch in the world. As you travel the roads you will see old farm houses and range cabins used by the ranch hands who worked for the Gang Ranch. The thunder eggs appear in different locations so keep your eyes open as you go along. This is wilderness so be sure your truck is in good working order and have a good spare, water, oil, etc. Also, be aware of the bears in the area and keep your food in the vehicle.

Thunder egg samples taken from the end of Perlite Road.

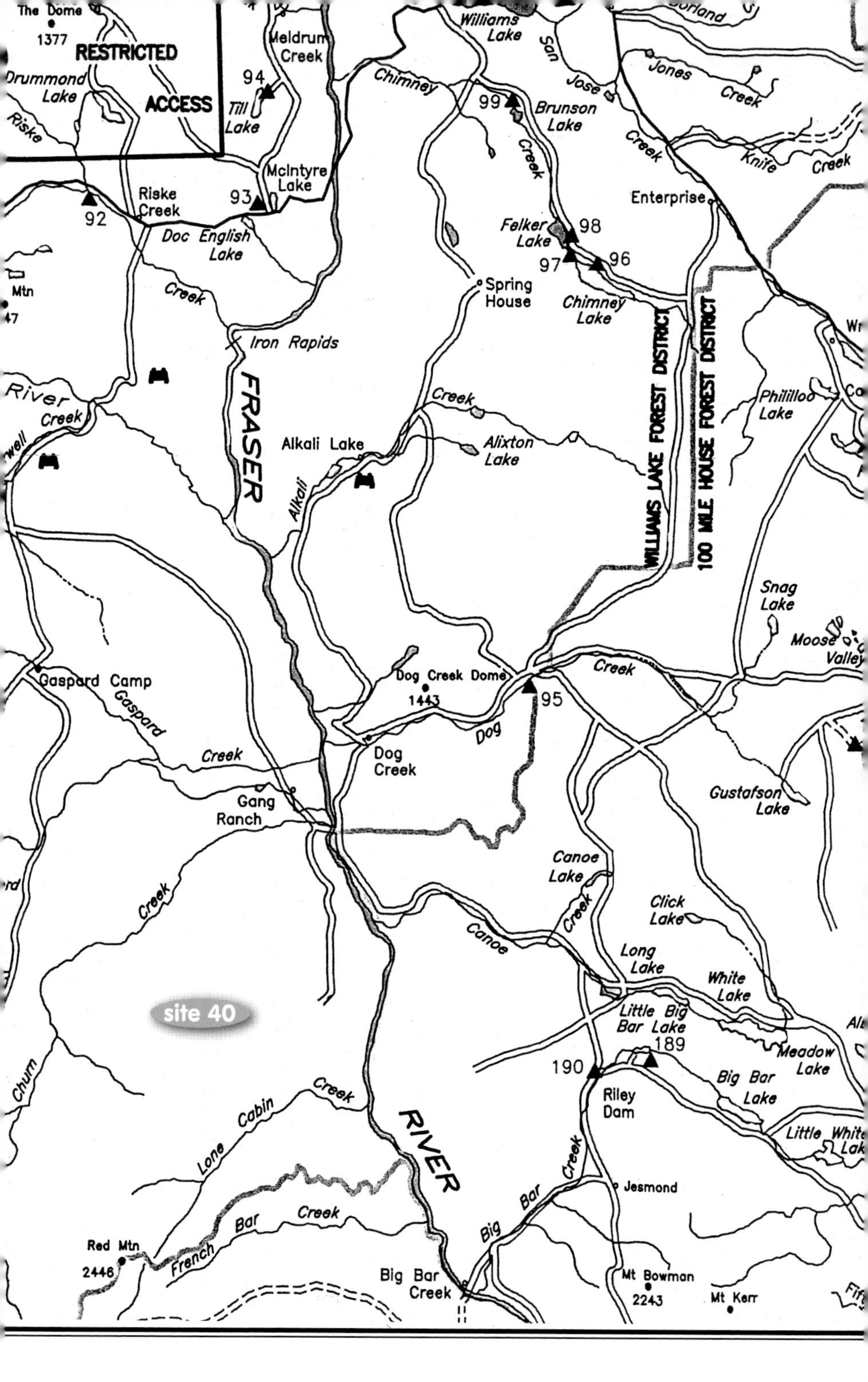

The Dome
1377
RESTRICTED
ACCESS
Drummond Lake
Riske
Meldrum Creek
94
Till Lake
McIntyre Lake
93
Riske Creek
92
Doc English Lake
Creek
Iron Rapids
FRASER
River
Creek
Alkali Lake
Alkali
Williams Lake
San Jose
Chimney
99
Brunson Lake
Creek
Jones Creek
Creek
Knife Creek
Enterprise
Felker Lake
98
97
96
Spring House
Chimney Lake
WILLIAMS LAKE FOREST DISTRICT
100 MILE HOUSE FOREST DISTRICT
Phililloo Lake
Creek
Alixton Lake
Snag Lake
Moose Valley
Dog Creek Dome
1443
95
Creek
Dog
Dog Creek
Gaspard Camp
Gaspard
Creek
Gang Ranch
Gustafson Lake
Creek
Canoe Lake
Creek
Canoe
Click Lake
Long Lake
White Lake
Little Big Bar Lake
189
190
Meadow Lake
Big Bar Lake
Riley Dam
Little White Lake
site 40
Churn
Creek
Lone Cabin
RIVER
Big Bar Creek
Jesmond
French Bar Creek
Red Mtn
2446
Big Bar Creek
Mt Bowman
2243
Mt Kerr

site 41 NEEDA LAKE

▸ map page 73

geodes

This location is north of Cache Creek on Highway 97 before the town of 100 Mile House. On Highway 97, 11 km before 100 Mile House, turn east on Highway 24. In approximately 36 km, turn left on Horse Lake Road. The turnoff has a small strip mall at the northeast corner. At 7.5 km turn right on Judson Road. Proceed to Needa Lake turnoff at 9.2 km. Turn left and proceed 6.2 km to the north side of the lake. There you will see the geodes in the basalt on the left. Some of these geodes can get quite large, and the best way to find them is with sledge hammers and chisels. When you come across one, you must be very careful on your approach, as they will break quite easily. Be sure to keep the road clean. There are wilderness camps on the lake that have tables and toilets. The lake is very good for fishing most of the year.

Inside one of the Needa Lake geodes.

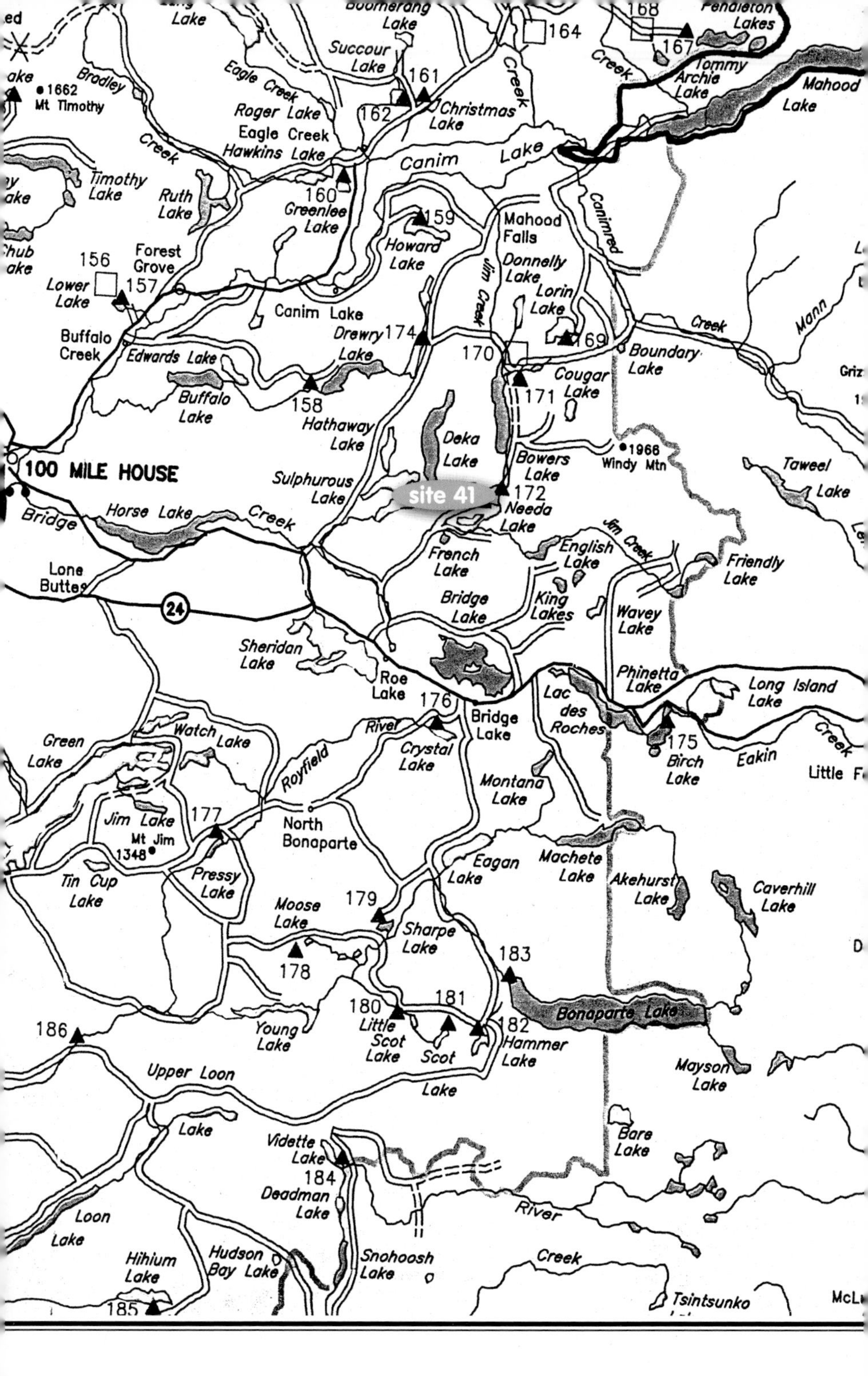

Boomerang Lake
Succour Lake
Pendleton Lakes
164
168
167
Tommy Archie Lake
Mahood Lake
Eagle Creek
Bradley Creek
1662 Mt Timothy
161
162
Christmas Lake
Creek
Roger Lake
Eagle Creek
Hawkins Lake
Canim Lake
Timothy Lake
Ruth Lake
160
Greenlee Lake
159
Howard Lake
Mahood Falls
Canimred
Donnelly Lake
Lorin Lake
Jim Creek
156
Forest Grove
Lower Lake
157
Canim Lake
Buffalo Creek
Drewry Lake
174
169
170
Creek
Mann
Boundary Lake
Edwards Lake
Buffalo Lake
158
Cougar Lake
171
Hathaway Lake
Deka Lake
Bowers Lake
1966 Windy Mtn
Taweel Lake
100 MILE HOUSE
Sulphurous Lake
site 41
172
Needa Lake
Bridge
Horse Lake
Creek
Jim Creek
French Lake
English Lake
Friendly Lake
Lone Butte
24
Bridge Lake
King Lakes
Wavey Lake
Sheridan Lake
Roe Lake
Phinetta Lake
Long Island Lake
176
Lac des Roches
Bridge Lake
175
Birch Lake
Eakin
Creek
Little F
Green Lake
Watch Lake
River
Royfield
Crystal Lake
Montana Lake
Jim Lake
Mt Jim
1348
177
North Bonaparte
Eagan Lake
Machete Lake
Akehurst Lake
Caverhill Lake
Tin Cup Lake
Pressy Lake
Moose Lake
179
Sharpe Lake
183
178
180
181
Bonaparte Lake
186
Young Lake
Little Scot Lake
Scot
182
Hammer Lake
Mayson Lake
Upper Loon
Lake
Lake
Vidette Lake
184
Bare Lake
Deadman Lake
River
Loon Lake
Hihium Lake
Hudson Bay Lake
Snohoosh Lake
Creek
185
Tsintsunko

site 42 TRANQUILLE

▸ map page 77

fossils

From the bridge between Kamloops and North Kamloops, travel on Fortune Drive which turns into Tranquille Highway. Follow the road beside the tracks for 17 km. Turn right on Red Lake Road. Park here, where the road starts up the long hill. Walk along the tracks for half a kilometre, and as you go you will notice a large draw. The fossils are near the end of the draw. There should be other locations for the fossils in this area, so be sure to check around. A good place to start your search will be walking along the tracks. Don't just watch the railroad cuts, also look up each draw as you go. There are many good forestry campsites as well as wilderness sites. To get to the best of these, continue along the road that you came in on, and you will find them all along the forestry roads ahead. Be sure to watch for cattle and logging trucks when picking your campsite.

Park on Red Lake Road before the long hill and walk down the train tracks.

site 43 FREDERICK ROAD

▸ map page 77

agate

From the bridge between Kamloops and North Kamloops, travel on Fortune Drive which turns into Tranquille Highway. Turn right on Red Lake Road and proceed 18.9 km from the bridge to Frederick Road. From the turnoff, go about 457 metres and take the road leading off to the right. This road is closed to traffic, so park here and hike up the road for about a half a kilometre. The agates are in the hills on your right, and in veins and nodules throughout the small valley in front of you. There are many agates here, but a lot of them are a plain white. There are quite a few with some nice bands and these are the ones that are preferred. The larger agates are around two or three pounds. There are good campsites in the immediate area, but you will need to bring water. The large cliffs overlooking Kamloops Lake just past here are called Battle Rock. It was a major fortress for the local First Nations because of its ease of defense. Because of this, you can find arrowheads on the rock and in the surrounding area.

View of the hills containing the agate.

site 44 WATCHING CREEK

▸ map page 44

agate

From the bridge between Kamloops and North Kamloops, travel on Fortune Drive which turns into Tranquille Highway. Turn right on Red Lake Road and proceed 27 km from the bridge in Kamloops. Just before the 14-km forestry road marker, there is a sign indicating the Watching Creek forest campsite. Proceed to the campsite and at the north end follow the deactivated road. Walk down the hill and cross the river on the logs lying across it. The river is shallow, so if you don't like the look of the logs, you can take off your shoes and walk across. The agates are in this bank and in many other banks in this area. The agates here are quite numerous. If you climb this hill and cross the large field, you will be in another agate location, which also contains common opal.

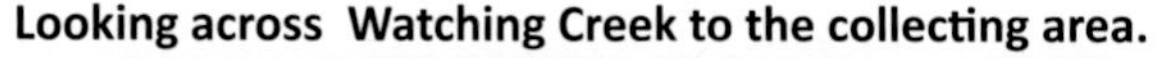

Looking across Watching Creek to the collecting area.

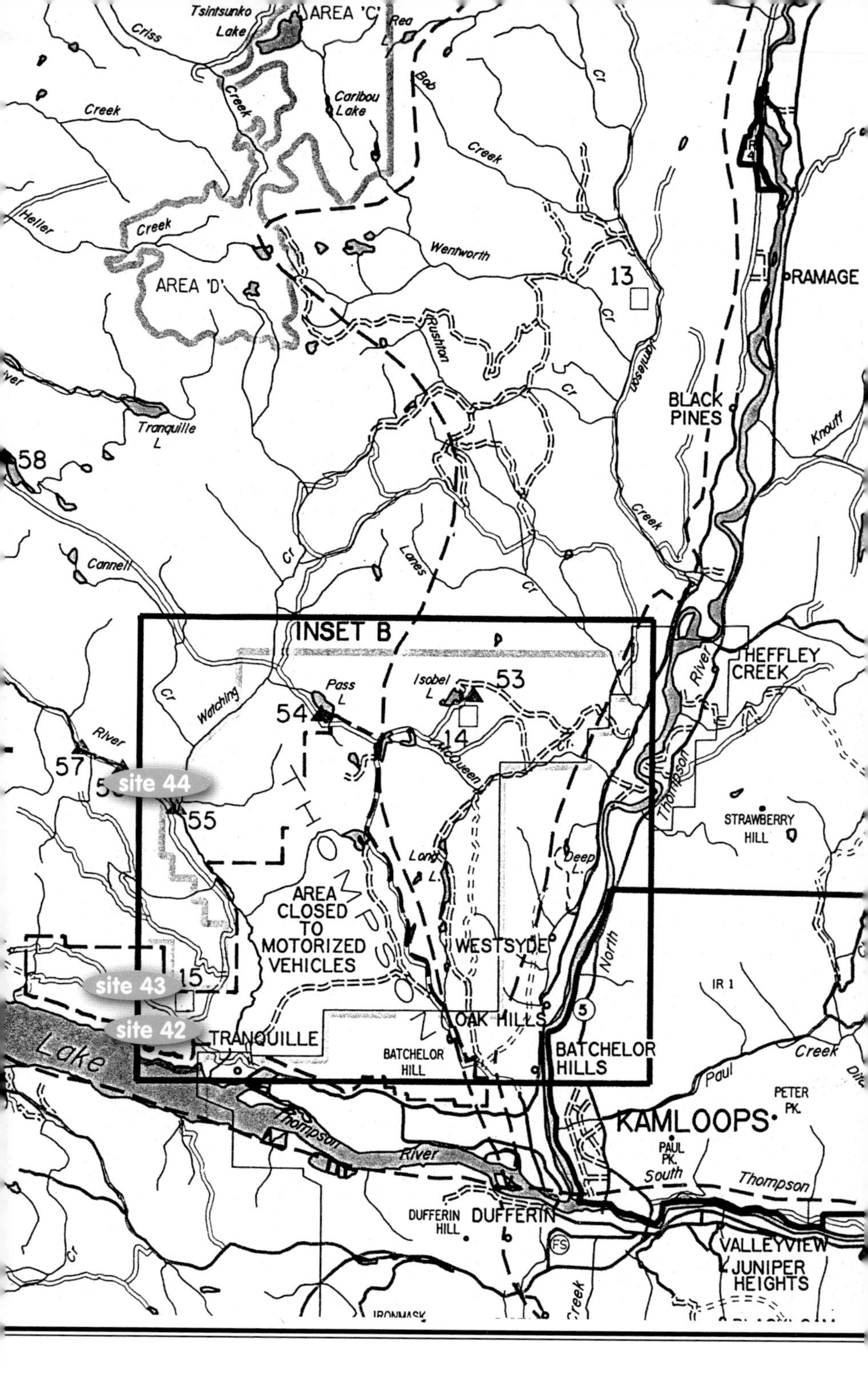

Tsintsunko Lake
AREA 'C'
Rea
Criss
Creek
Creek
Caribou Lake
Bob
Creek
Cr
Heller
Creek
Wentworth
AREA 'D'
13
RAMAGE
Cr
Rushton
Cr
Jamieson
Tranquille L
BLACK PINES
Knouff
58
Creek
Cannell
Cr
Lanes
Cr
INSET B
HEFFLEY CREEK
River
Pass L
Isobel L
53
Cr
Watching
54
14
McQueen
Cr.
River
57
site 44
Thompson
STRAWBERRY HILL
55
Long L.
Deep L.
THOMPSON
AREA CLOSED TO MOTORIZED VEHICLES
WESTSYDE
North
IR 1
site 43
15
site 42
OAK HILLS
5
TRANQUILLE
Lake
BATCHELOR HILL
BATCHELOR HILLS
Creek
Paul
PETER PK.
KAMLOOPS
Thompson
River
PAUL PK.
South
Thompson
DUFFERIN HILL
DUFFERIN
FS
VALLEYVIEW
JUNIPER HEIGHTS
Cr
Creek

site 45 ADAMS LAKE

calcite

From Kamloops, take Highway 5 for 55 km to the Agate Bay exit. Proceed for about 27 km and turn left at the stop sign. From the stop sign, travel 10 km until you reach the 29-km marker, where, on your left there is a large rock face. Near the base of the rock-face, where the trees end, approximately in the middle, you will find a cave that has a lot of calcite in it. We have found calcite plates that we had to leave behind because they were too heavy to carry. The terrain above the road to the calcite is quite steep. There are forestry campsites on the lake as well as good camping spots along the forestry roads in the area. This lake is fed, as well as drained by, the Adams River, which is famous for the salmon runs that occur there. There are sockeye, chinook, coho and pink salmon. Spawning beds are located in the Roderick Haig-Brown Park, which is located along both sides of the Adams River from Adams Lake.

Calcite crystal.

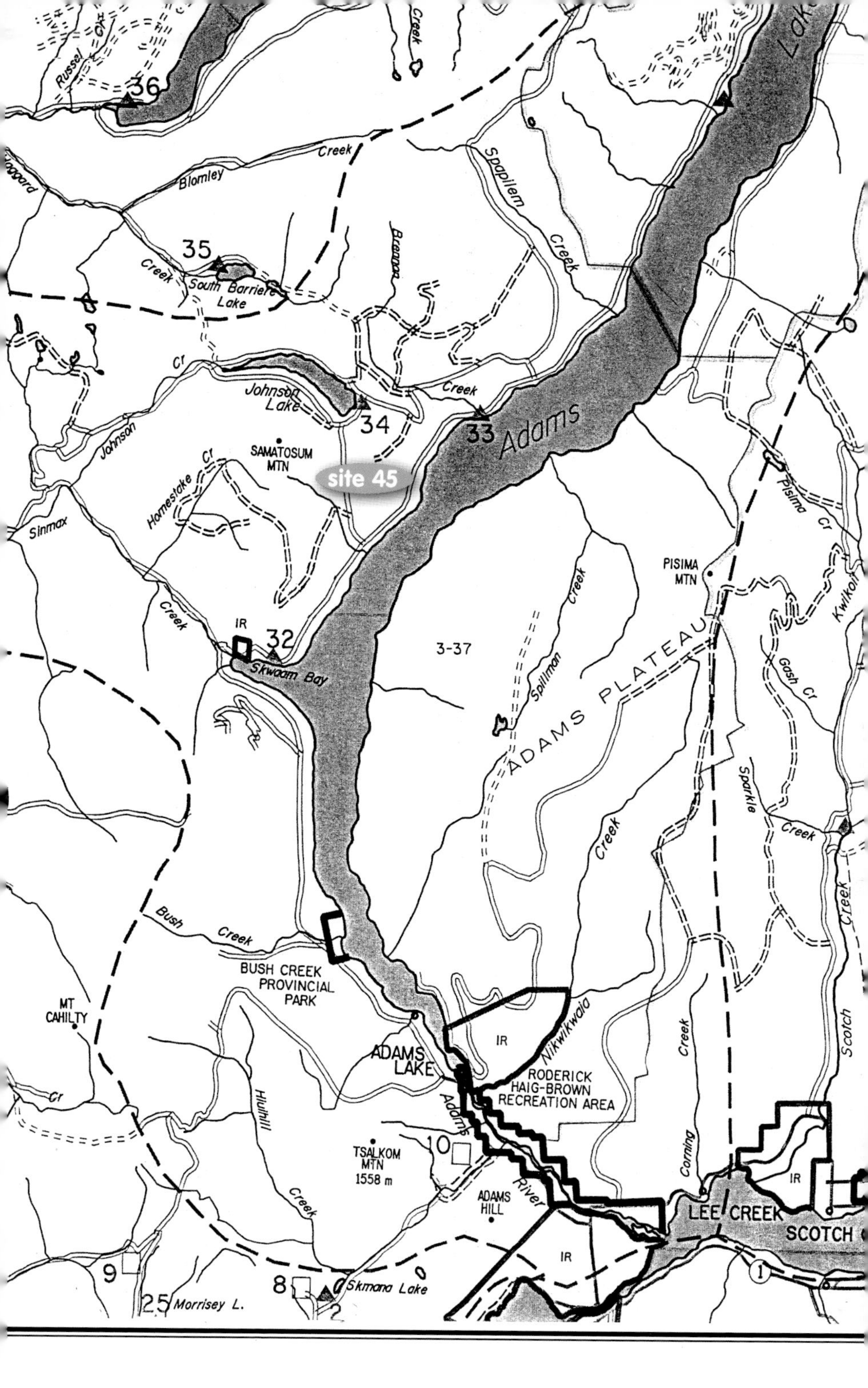
36
35
34
33
32
Russel Cr
Creek
Blomley
Creek
South Barriere Lake
Cr
Johnson Lake
Creek
Johnson
SAMATOSUM MTN
site 45
Homestake Cr
Sinmax
Creek
IR
Skwaam Bay
Adams
Lake
Spapilem
Creek
Brennan
3-37
Spillman
Creek
ADAMS PLATEAU
PISIMA MTN
Pisima Cr
Gash Cr
Kwikoit
Sparkle
Creek
Creek
Bush
Creek
BUSH CREEK PROVINCIAL PARK
MT CAHILTY
Cr
ADAMS LAKE
IR
Nikwikwaia
RODERICK HAIG-BROWN RECREATION AREA
Adams
River
Hiuihill
Creek
TSALKOM MTN 1558 m
ADAMS HILL
10
IR
Creek
Corning
Scotch
Creek
IR
LEE CREEK
SCOTCH
1
9
8
2
25
Morrisey L.
Skmana Lake

site 46 JOSEPH CREEK

▸ map page 81

fossils

From Kamloops head north on Highway 5 into Little Fort. Cross on the ferry at Little Fort, and drive for 13 km to Joseph Creek. Park here and follow the turnoff road to the rock bluff. The fossils are at the bottom of the bluff and in the talus. The last time I was there, it had been recently blasted, so you really need to search for the material. If you spend a little time, you will find the fossils and possibly a new vein. Look at the pieces with fossils in them and search for the host rock. There are many good campsites in the area with toilets and tables as well as many good wilderness spots. The fishing is good in the lakes, but seems to be best in the summer. The roads can be slippery after a rain, so be careful of other traffic.

site 47 BARRIERE

▸ map page 81

jasper

The Barriere River contains some nice jasper as well as other collectables. You can get to the river at the highway bridge.

site 48 RED RIDGE

▸ map page 81

fluorite

There is fluorite of many color varieties and at frequent intervals on the summit of Red Ridge. The local clubs have had trips here and the collecting was good. Get in touch with the clubs in the area for more information on the location. This is on the south side of the Thompson River. It is southeast of Birch Island Station. I have seen some nice samples of the fluorite; they were of a good size and a deep purple.

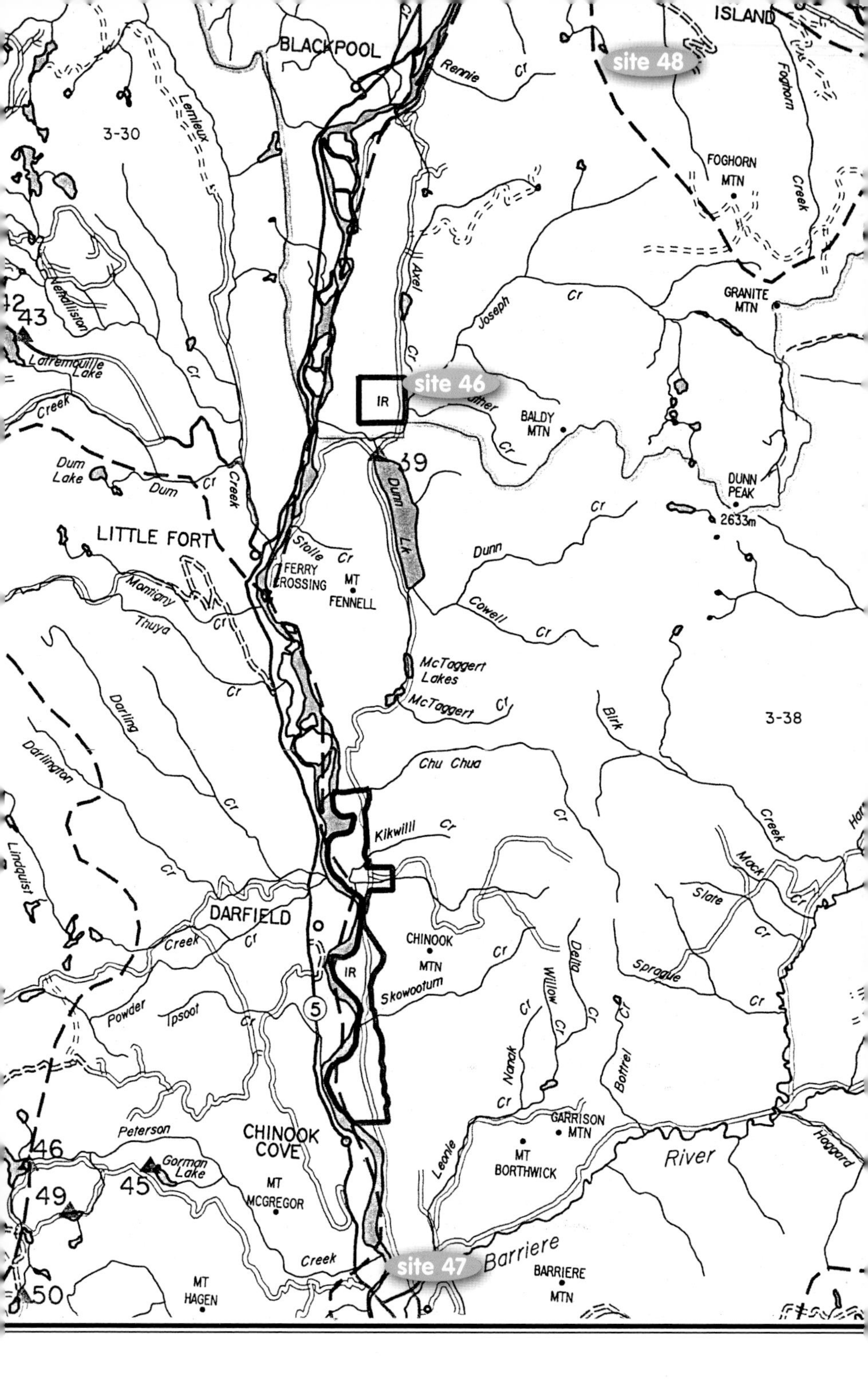

BLACKPOOL
ISLAND
site 48
Rennie Cr
Foghorn Creek
FOGHORN MTN
3-30
Lemieux
Axel Cr
Joseph Cr
GRANITE MTN
43
Latremouille Lake
Creek
site 46
IR
BALDY MTN
Cr
39
Dum Lake
Dum Cr
Creek
Dunn Lk
DUNN PEAK
2633m
LITTLE FORT
Stolle Cr
FERRY CROSSING
MT FENNELL
Dunn
Cowell Cr
Montigny
Thuya Cr
Cr
McTaggert Lakes
McTaggert Cr
Darling
Darlington
Cr
Birk
3-38
Chu Chua
Cr
Kikwilli Cr
Creek
Lindquist
Cr
Mack
Slate
Cr
DARFIELD
Creek
Cr
CHINOOK MTN
Cr
Delta
Willow Cr
Sprague
Cr
IR
Skowootum
5
Powder
Tpsoot
Cr
Cr
Nanak
Cr
Bottrel
Cr
GARRISON MTN
Peterson
CHINOOK COVE
46
Gorman Lake
45
49
MT MCGREGOR
Leonie
MT BORTHWICK
River
Haggard
Creek
site 47
Barriere
BARRIERE MTN
MT HAGEN
50

site 49 CHEZACUT

▸ map page 83

opal

From the town of Williams Lake take Highway 20 for 145.5 km where you will turn right on Chezacut forest service road. This area is remote so go past the road and gas up at the Redstone gas station which is about 20 km from the old town of Redstone. The best landmark is the cement bridge; the Chezacut Road is just east of that. Go up the Chezacut forest service road for 52 km and turn left on road 147. It is at the 147-km marker and is locally called Chilcotin or Red Top Road. Follow this road across the creek and up the hill for 8 km. The agate and opal are at the crest of the hill in crumbling basalt. There should be more in the surrounding hills.

Collecting agate and opal.

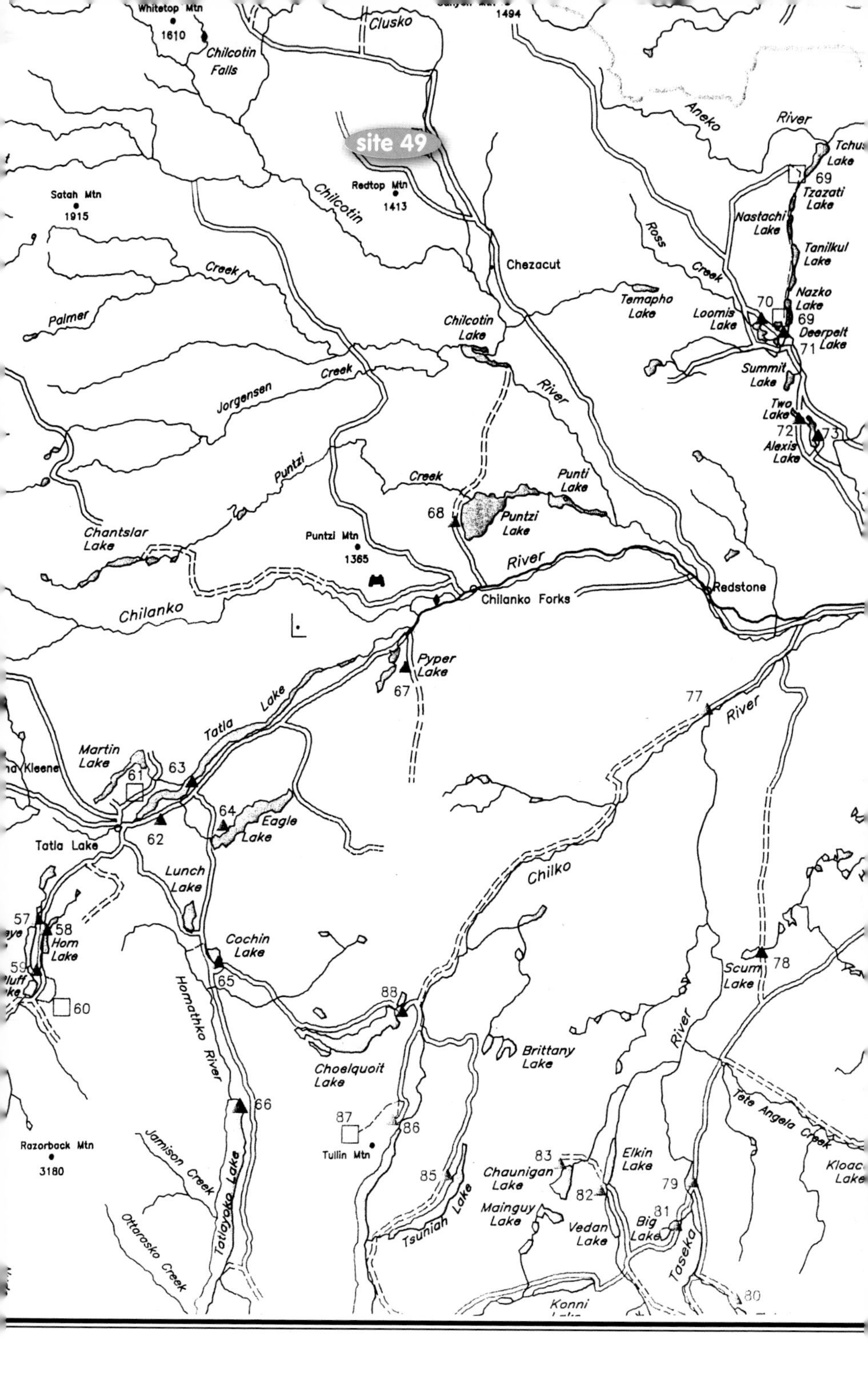

Whitetop Mtn
1610
Chilcotin
Falls
Clusko
1494
Aneko
River
site 49
Redtop Mtn
1413
Chilcotin
Satah Mtn
1915
Tchus
Lake
69
Tzazati
Lake
Nastachi
Lake
Ross
Tanilkul
Lake
Chezacut
Creek
Creek
Nazko
Lake
Temapho
Lake
70
Loomis
Lake
69
Palmer
Chilcotin
Lake
Deerpelt
Lake
71
Summit
Lake
Jorgensen
Creek
River
Two
Lake
72
73
Alexis
Lake
Puntzi
Creek
Punti
Lake
68
Puntzi
Lake
Chantslar
Lake
Puntzi Mtn
1365
River
Redstone
Chilanko
Chilanko Forks
Pyper
Lake
67
77
River
Tatla
Lake
Martin
Lake
Kleene
63
61
64
Eagle
Lake
62
Tatla Lake
Chilko
Lunch
Lake
57
58
Horn
Lake
Cochin
Lake
Scum
Lake
78
65
60
88
Homathko River
River
Brittany
Lake
Choelquoit
Lake
Tete Angela Creek
66
87
86
Razorback Mtn
3180
Tullin Mtn
Jamison Creek
Elkin
Lake
Kloac
Lake
83
Chaunigan
Lake
85
Tatlayoko Lake
79
82
Mainguy
Lake
81
Big
Lake
Vedan
Lake
Tsuniah Lake
Ottarosko Creek
Taseka
80
Konni

site 50 WELLS

▸ map page 85

pyrite

The town of Wells has some nice samples of pyrite in the mine dumps around town. To get to the town of Wells head out of Quesnel on Highway 26E for 80.7 km. As you approach the town you will see the large mine dumps to the right of the town. Here you will also find some core samples and peacock ore. This historic area will have many mines in the area, but there are also many claims. Be sure the area where you are collecting is open. In the area around Cottonwood, we found some quartz crystals, but when we returned to the area the road we found them on had changed. There is a lot of logging in the area, so be sure to check all quartz veins.

One of the mine dumps around Wells.

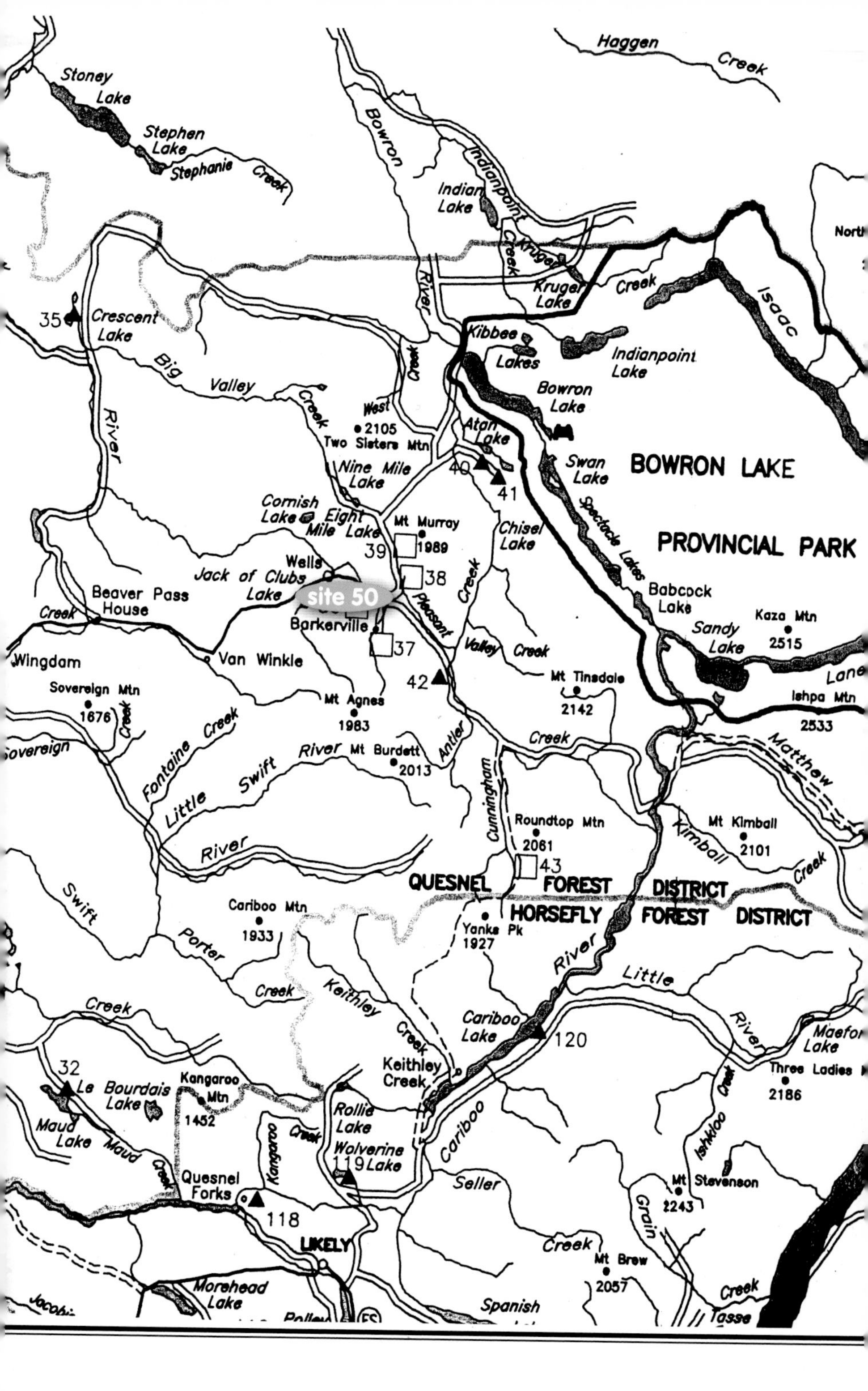
Haggen
Creek
Stoney
Lake
Stephen
Lake
Stephanie
Creek
Bowron
River
Indianpoint
Creek
Indian
Lake
Kruger
Kruger
Lake
Creek
Isaac
North
35
Crescent
Lake
Big
Valley
Creek
River
Kibbee
Lakes
Indianpoint
Lake
Bowron
Lake
West
Creek
2105
Two Sisters Mtn
Atan
Lake
Swan
Lake
BOWRON LAKE
PROVINCIAL PARK
Nine Mile
Lake
40
41
Spectacle Lakes
Cornish
Lake
Eight
Mile Lake
Mt Murray
1989
39
38
Chisel
Lake
Wells
Jack of Clubs
Lake
site 50
Beaver Pass
House
Creek
Barkerville
Pleasant
Creek
Babcock
Lake
Kaza Mtn
2515
Sandy
Lake
Wingdam
Van Winkle
37
Valley
Creek
42
Mt Tinsdale
2142
Lane
Ishpa Mtn
2533
Sovereign Mtn
1676
Creek
Mt Agnes
1983
Sovereign
Fontaine Creek
Little
Swift
River
Mt Burdett
2013
Antler
Creek
Matthew
Cunningham
Roundtop Mtn
2061
Mt Kimball
2101
Kimball
Creek
River
43
QUESNEL
FOREST
DISTRICT
HORSEFLY
FOREST
DISTRICT
Swift
Cariboo Mtn
1933
Yanks Pk
1927
Porter
Creek
River
Little
Keithley
Creek
Creek
Cariboo
Lake
120
River
Maefor
Lake
Keithley
Creek
32
Le Bourdais
Lake
Kangaroo
Mtn
1452
Three Ladies
2186
Maud
Lake
Maud
Creek
Kangaroo
Creek
Rollie
Lake
Wolverine
119
Lake
Cariboo
Ishkloo
Creek
Quesnel
Forks
118
Seller
Mt
2243
Stevenson
Grain
LIKELY
Creek
Mt Brew
2057
Morehead
Lake
Jacobi
Spanish
Creek
Tasse
FS

site 51 MARLIN ROAD ▸ map page 87

agate, calcite

From Highway 16 at the Augier Road junction you will turn right if you are coming from Fraser Lake and left if you are coming from Burns Lake. Augier Road junction is 22.8 km east of Burns Lake. There is a flashing yellow light at the intersection. Just before the 7-km sign you will turn right on Hannay Road. At 35.7 km you will turn onto Pearson Road. Keep left at the fork, and at 3.3 km turn on Marlin Road. The agates are in the last 3 km of the road itself and over the small banks on both sides of the road. Always check the new logging shows for new areas, as each of the roads mentioned here have produced agates.

Agates found in the Hannay Road area.

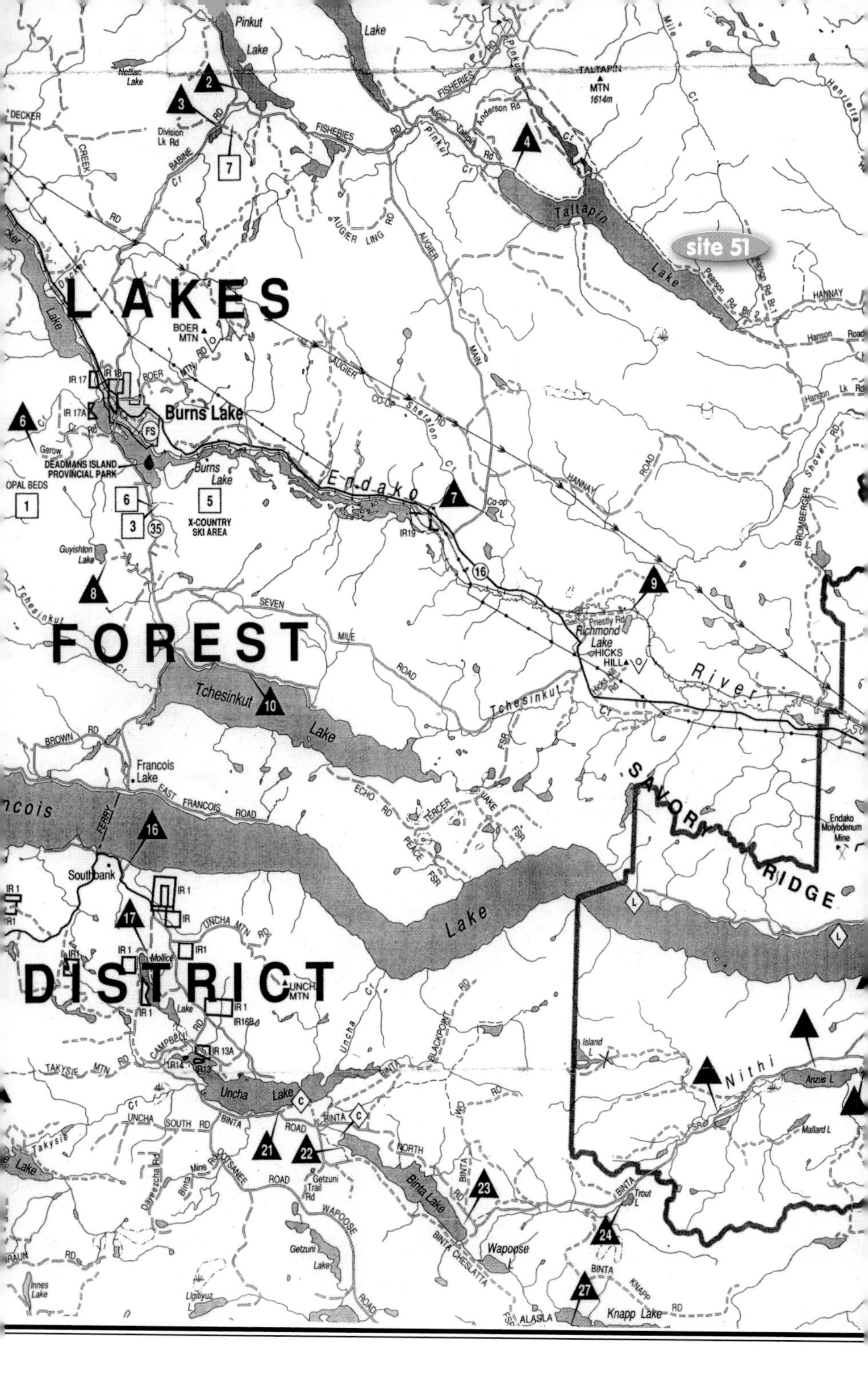

LAKES
FOREST
DISTRICT
Pinkut Lake
Taltapin Lake
TALTAPIN MTN 1614m
site 51
BOER MTN
Burns Lake
DEADMANS ISLAND PROVINCIAL PARK
OPAL BEDS
X-COUNTRY SKI AREA
Guyishton Lake
Tchesinkut Lake
Francois Lake
Southbank
Richmond Lake
HICKS HILL
Endako
Endako Molybdenum Mine
SAVORY RIDGE
UNCHA MTN
Uncha Lake
Binta Lake
Wapoose L
Knapp Lake
Getzuni Lake
Innes Lake
Island L
Nithi
Anzus L
Mallard L
Trout L

site 52 SPRINGER CREEK

▸ map page 89

quartz crystals

Across Highway 6 from Slocan you will find Springer Creek forest service road. At 2.5 km from the highway on Springer Creek Road, you will turn right on Ottawa Hill forest service road. From here you will travel to the 6.5-km marker and turn right on Dayton forest service road. At approximately the 7.6-km marker you will find quartz crystals in the road cut. If you carry on to just before the 9-km marker you will find small quartz crystals and pyrite.

site 53 METEOR MINE

▸ map page 89

silver, core samples

Looking down from the main mine.

Across Highway 6 from Slocan you will find Springer Creek forest service road. At 2.5 km from the highway on Springer Creek road, turn right on Ottawa Hill forest service road. Follow this road to the 12.5 marker and turn right on the Meteor Mine road. Follow this to the end, where you will find the old Meteor Mine. The last part of this road you will need a 4-wheel-drive vehicle. Check the mine dumps for lots of pyrites and silver minerals.

site 54 ARLINGTON MINE

▸ map page 89

silver minerals

Across Highway 6 from Slocan you will find Springer creek forest service road. At 2.5 km from the highway on Springer Creek road, you will turn right on Ottawa Hill forest service road. Follow this road to the 12.5-km marker. Keep to the left and head

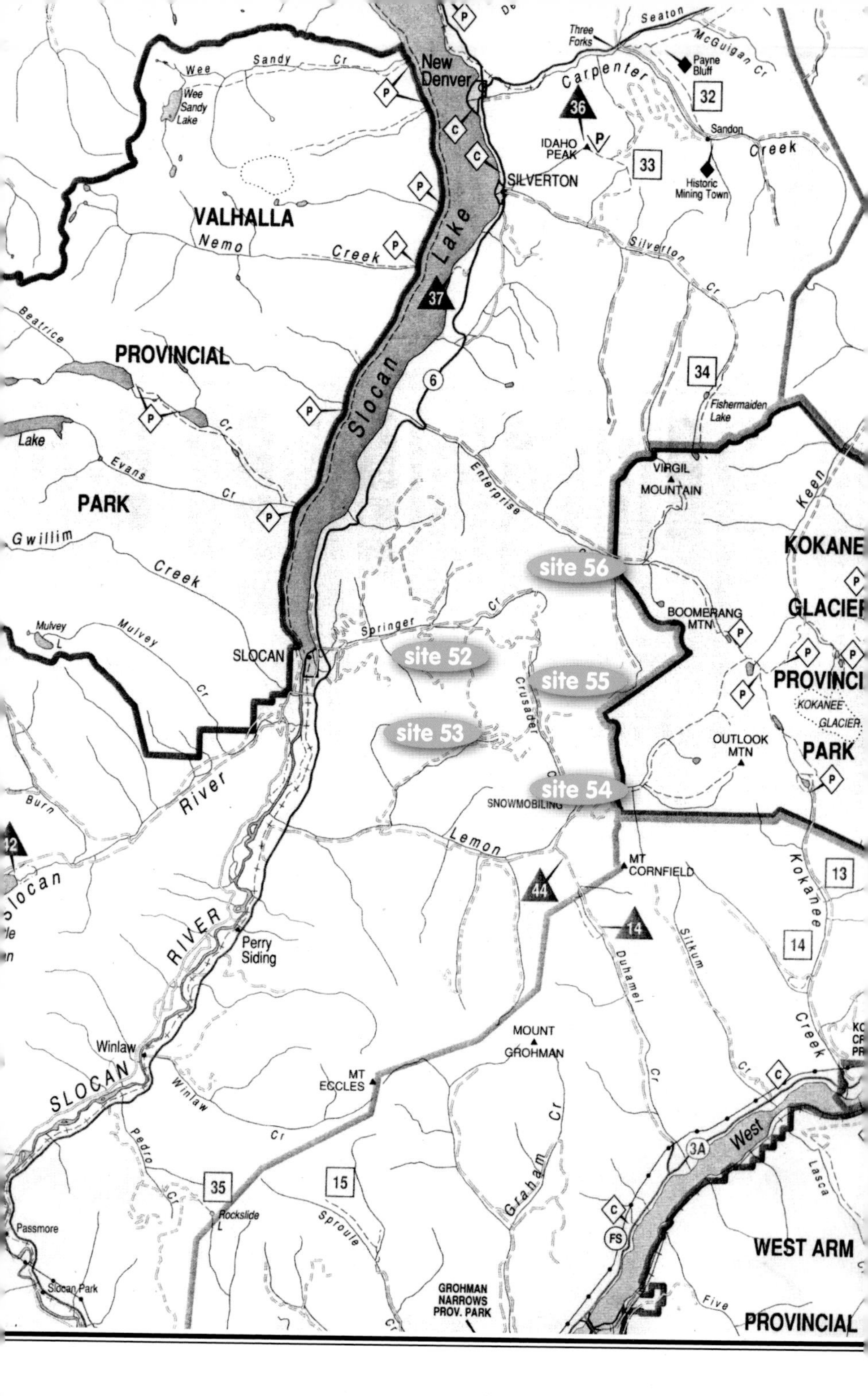
New Denver
Three Forks
Seaton
McGuigan Cr
Payne Bluff
Carpenter
32
36
IDAHO PEAK
Sandon
Creek
33
SILVERTON
Historic Mining Town
Wee Sandy Cr
Wee Sandy Lake
VALHALLA
Nemo Creek
Slocan Lake
37
Silverton Cr
Beatrice
PROVINCIAL
34
6
Fishermaiden Lake
Lake
Cr
Evans Cr
VIRGIL MOUNTAIN
Enterprise
Keen
PARK
Gwillim Creek
site 56
KOKANEE
GLACIER
BOOMERANG MTN
Mulvey L
Mulvey Cr
Springer Cr
SLOCAN
site 52
site 55
PROVINCIAL
Crusader
KOKANEE GLACIER
site 53
OUTLOOK MTN
PARK
site 54
SNOWMOBILING
Burn
River
Lemon
MT CORNFIELD
13
Slocan
44
14
Kokanee
RIVER
Perry Siding
Silkum
Duhamel
14
MOUNT GROHMAN
Creek
Winlaw
MT ECCLES
Cr
SLOCAN
Winlaw
Cr
Cr
Graham Cr
3A
West
Pedro
Cr
35
15
Lasca
Rockslide L
Sproule
FS
Passmore
WEST ARM
Slocan Park
GROHMAN NARROWS PROV. PARK
Cr
Five
PROVINCIAL

down and cross the creek. Around the 14-km marker you will find Arlington Mine road heading off to the right. At the junction you will see the remains of the old mine. Search here for silver minerals and on the lower section for core samples.

Mine dump at Arlington Mine.

site 55 LITTLE TIM MINE

▸ map page 89

silver minerals

From the town of Slocan, across Highway 6, take the Springer Creek forest service road. Follow this main road and at 10 km keep to the right. After 14.8 km turn to the right. Follow this road to the end which is at 21 km. This is the Little Tim Mine. Check the dumps for galena and other silver ores. You can also find silver. Some wire silver has been found in the small quartz crystal vugs.

site 56 ENTERPRISE MINE

▸ map page 89

silver and copper minerals

Mine dump at Enterprise Mine.

From the town of Slocan, head north on Highway 6 for 15 km. Cross the Enterprise Creek bridge and carry on for .7 km where you will turn right on the Enterprise Creek road. Travel on this road for 7 km where you will reach a bridge on your right. You will be crossing this bridge, but before you do carry on for a little ways where you will see the mine on your right, across the creek. Get your bearings and go back to the bridge and follow the logging roads to get as close as you can to the mine.

aragonite

From the town of Trail, head out Highway 3b and turn right at the Highway 22a exit. This is 6.7 km from the bridge in town. From here turn left at 8.3 km on to 7 Mile Dam road. At 8 km, on the left-hand side of the roadcut you will find aragonite. The surrounding rock is quite crumbly so pry off a large section and break it up with a hammer. The crystals are fragile and come in large veins and in vugs.

Collecting aragonite beside the road.

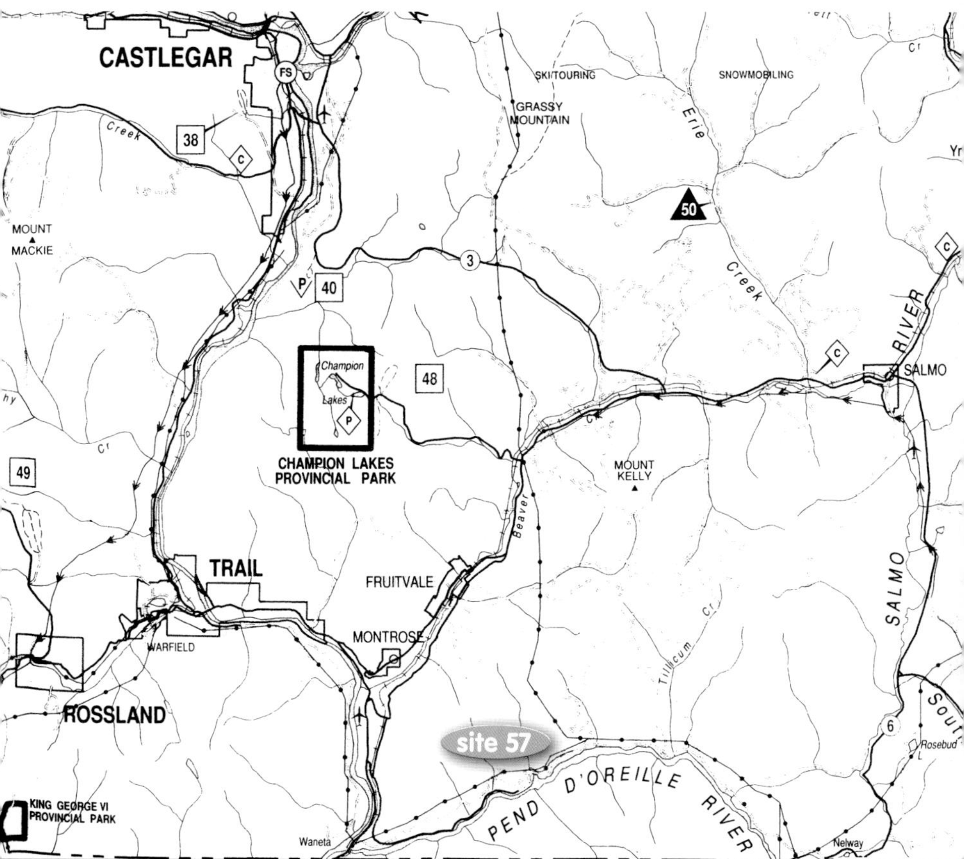

HODER CREEK

▸ map page 93

garnet, tourmaline

From the town of Slocan, heading south on Highway 6, turn right on Gravel Pit road. The beginning of the road is the 2-km forest service marker. Just after the 4-km marker, keep right on Little Slocan forest service road. At the 15-km marker, go straight.Take the Hoder Creek forest service road 1 km past the 21-km marker. The markers start over at 0. Just before the 6-km marker, you will keep to the left, and at .8 of a km, you will be crossing the Berry Creek bridge. From the bridge, it is 1.5 km to the site. The garnet is in the black granite, and the tourmaline is right beside it. The tourmaline is for displays and may not take a polish.

Parked near the garnets.

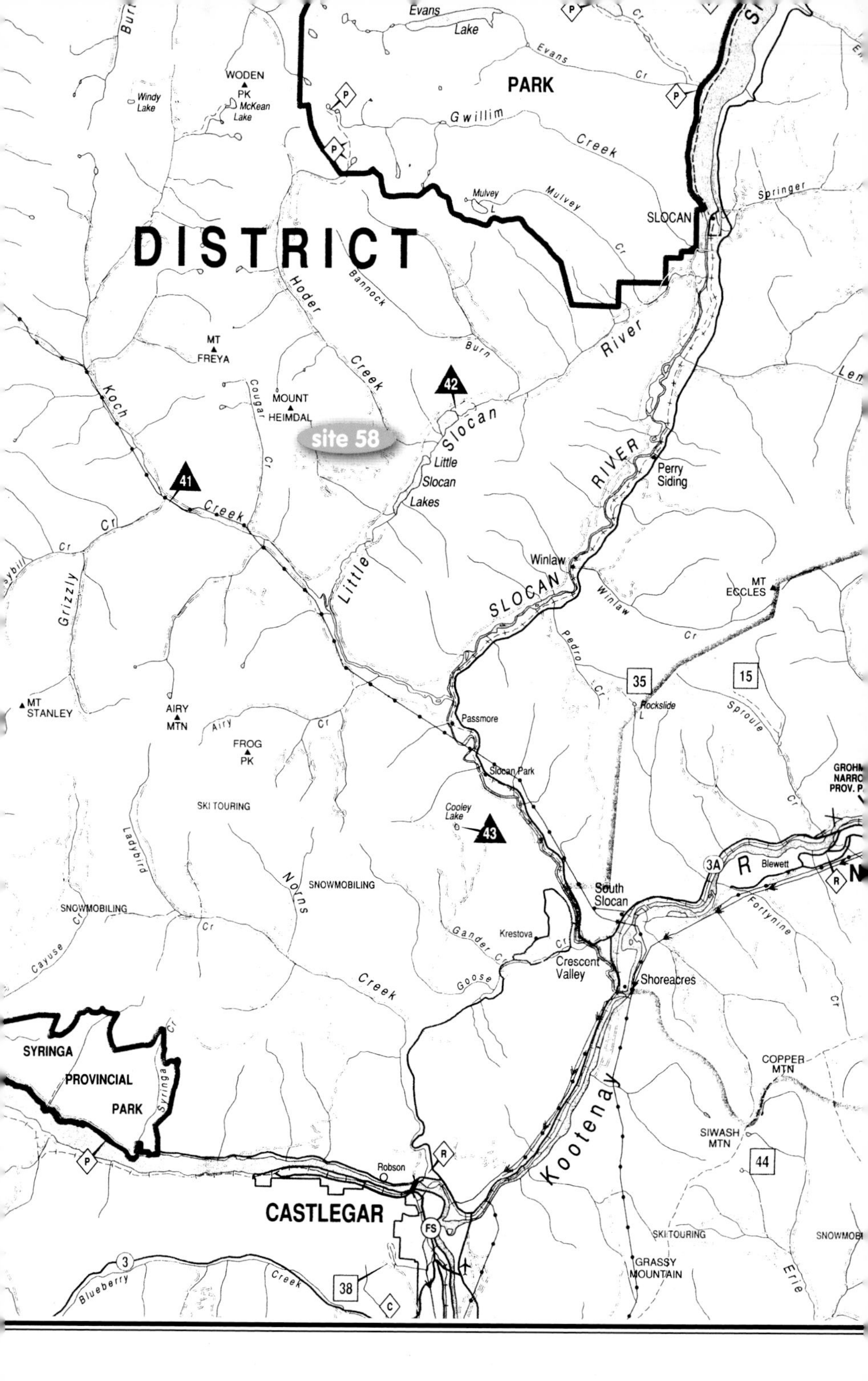
DISTRICT
PARK
Evans Lake
Evans Cr
Gwillim Creek
Mulvey L
Mulvey Cr
SLOCAN
Springer
WODEN PK
Windy Lake
McKean Lake
Hoder Creek
Bannock Burn
Slocan River
MT FREYA
MOUNT HEIMDAL
Cougar Cr
Koch Creek
site 58
Little Slocan Lakes
Little Slocan
RIVER
Perry Siding
Winlaw
SLOCAN
Winlaw Cr
MT ECCLES
Grizzly Cr
Pedro Cr
Rockslide L
Sproule Cr
MT STANLEY
AIRY MTN
Airy Cr
FROG PK
Passmore
Slocan Park
SKI TOURING
Cooley Lake
Ladybird
Norns Creek
SNOWMOBILING
Blewett
Fortynine
South Slocan
Krestova
Gander Cr
Crescent Valley
Goose
Shoreacres
Cayuse Cr
SYRINGA PROVINCIAL PARK
Syringa Cr
Kootenay
COPPER MTN
SIWASH MTN
Robson
CASTLEGAR
SKI TOURING
GRASSY MOUNTAIN
Blueberry Creek
Erie
SNOWMOBI
GROHM NARRO PROV. P

CLUBS & CONTACTS

1120 Rock Club
Dave Barclay > davebarclay@telus.net

Abbotsford Rock and Gem Club
Georgina Selinger > abbyrockngem@hotmail.com

Alberni Valley Rock and Gem Club
Dave & Dot West > compudoc@telus.net

B.C. Faceters Guild
Rob Giesbrecht > robgies@hotmail.com

Burnaby Laphounds Club
Nancy Dickson > nancyandallen@telus.net

Courtenay Gem and Mineral Club
Jan Boyes > janboyes@telus.net

Cowichan Valley Rockhound Club
Jennifer Proctor > jeproctor@sha2.ca

Creative Jewellers Guild
Ken McIntosh > ks.mcintosh@dccnet.com

Creston Valley Prospectors
Eric Kutzner > ekutzner@telus.netpowel

Delta Rockhounds Rock and Gem Club
Mary Cool > coolgirl@dccnet.com

Fraser Valley Rock and Gem Club
Karen Archibald > karchiba@telus.net

Golden Rock and Fossil Club
Stan Walker > swalker@persona.ca

Hastings Centre Rockhounds Club
Michael Edwards > anwards@shaw.ca

High Country Rockhound Club
Jan Kohar > gkohar@telus.net

Interlakes Rockhounders
Gary Babcock > greem@shaw.ca

Kokanee Rock Club
Bob Lerch > orgmec@shawbiz.ca

Lakes District Rock and Gem Club
Helen Brown > rkbrown49@hotmail.com

Maple Ridge Lapidary Club
Carol Kostachuck > caroloncall@shaw.ca

Parksville and District Rock and Gem Club
Marion Barclay > marybar@telus.net

Penticton Geology and Lapidary Club
Gloria Bordass

Port Moody Rock and Gem Club
Lisa Elser > lelser@mac.com

Powell River Lapidary Club
Jim Auchterlonie > achto@hotmail.com

Princeton Rock and Fossil Club
Franz Hofer > milo44@telus.net

Raft River Rockhounds
Fay McCracken > raftriverrockhounds@gmail.com

Richmond Gem and Mineral Club
Darlene Howe > darhowe@telus.net

Ripple Rock Gem and Mineral Club
Gordon Burkholder > jangor@telus.net

Selkirk Rock and Mineral Club
Maureen Krohman > mkrohman@shaw.ca

Shuswap Rock Club
Susan McLellan > shuswaprockclub@gmail.com

Spruce City Rock and Gem Club
Russ Davis > rsdavis@shaw.ca

Surrey Rockhound Club
Adam Steggings > rckhound@telus.net

Thompson Valley Rock Club
Helen Lowndes > helow@telus.net

Vernon Lapidary and Mineral Club
Pat O'Brien > patandphyl@telus.net

Victoria Lapidary and Mineral Society
Cameron Speedie > cspeedie@telus.net

INDEX

A

Agate: site 23, Boston Bar.

Ammonite fossil: site 5, Hale Creek/ Harrison Lake.

Aragonite: site 57, Trail.

Axinite, epidote, quartz crystal: site 13, Hope.

Calcite crystals: site 45, Kamloops.

Copper minerals/pyrites: sites 21, 50, 52, 53, 54, 55, 56.

Epidote and quartz crystal: site 12, Hope.

F

G

Fossil, clam: site 8, Mystery Creek.

Crinoid fossil: Chilliwack Lake Road.

Garnets: site 9, East Harrison Lake.

Geode, crystal-filled: site 41, 100 Mile House.

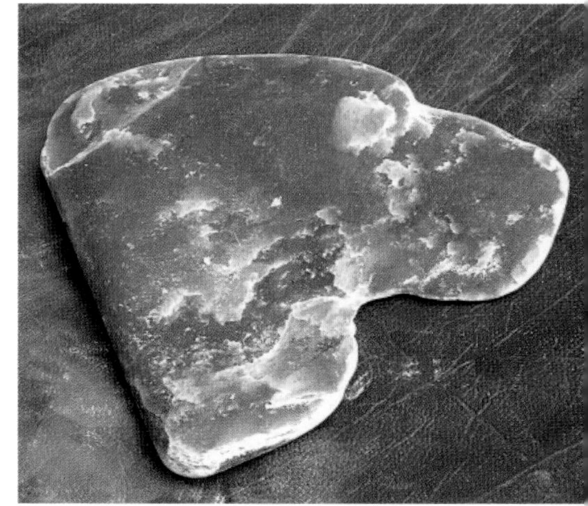

Jade: sites 2, 3, 4, 20, 22, Fraser River.

Petrified wood: site 28, Cache Creek.

Q

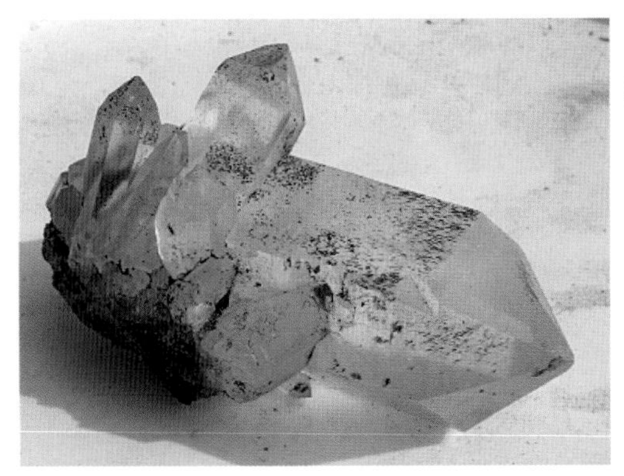

Quartz crystal: site 1, Whistler.

Chlorite-covered quartz crystal: site 1, Whistler.

Quartz crystal: site 1, Whistler.

Quartz crystal: site 12, Hope hydro station.

R

S

T

Z

Serpentine: sites 2, 3, 20, Hope.

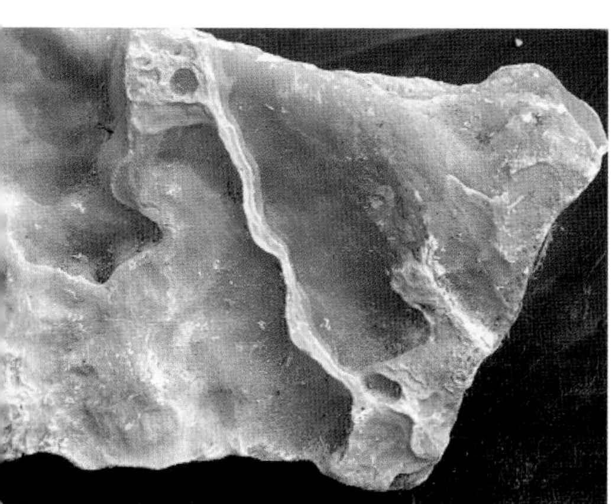

Travertine: site 22, Boston Bar.

Zeolite: Burns Lake area.

FIELD NOTES

FIELD NOTES